Changing Large Technical Systems

Changing Large Technical Systems

EDITED BY

Jane Summerton

Routledge
Taylor & Francis Group

NEW YORK AND LONDON

First published 1994 by Westview Press, Inc.

Published 2021 by Routledge
605 Third Avenue, New York, NY 10017
2 Park Square, Milton Park, Abingdon, Oxon OX14 4RN

Routledge is an imprint of the Taylor & Francis Group, an informa business

Library of Congress Cataloging-in-Publication Data
Changing large technical systems / edited by Jane Summerton.
    p. cm.
Papers presented at a roundtable conference held in Vadstena, Sweden in Aug. 1992.
   Includes bibliographical references and index.
   ISBN 0-8133-8817-1
   1. Technological innovations—Social aspects—Congresses.
I. Summerton, Jane.
T173.8.C37   1994
303.48'3—dc20   94-20880
                                                                            CIP

ISBN 13: 978-0-3670-1671-5 (hbk)
ISBN 13: 978-0-3671-6658-8 (pbk)

Contents

Acknowledgements

This book evolved from an idea to a research volume with the support of several colleagues and funding institutions.

The various chapters were originally presented at a roundtable conference on large technical systems in reconfiguration held in Vadstena, Sweden, in August 1992. The intellectual agenda of the conference was planned by an international scientific committee consisting of Arne Kaijser, Royal Institute of Technology in Stockholm; Bernward Joerges, Wissenschaftszentrum Berlin für Sozialforschung; Donald MacKenzie, University of Edinburgh; Svante Beckman, Linköping University; and myself. Following the conference, the committee also selected the chapters for this book.

Lars Ingelstam was an important source of support in the project. Anna Ghannadan provided crucial help in the endless practicalities of carrying out a conference. I am also indebted to Jennifer Nelson, who provided valuable editorial assistance and compiled the index, and to Barbro Axelsson, who formatted and typeset the book into camera-ready publishable form.

This project was made possible by the generous financial support of four major Swedish institutions for the support of social science and technical research. These are the Energy Systems Studies (AES) unit of the Swedish National Board for Industrial and Technical Research (NUTEK), the Bank of Sweden Tercentenary Foundation, the Swedish Council for Planning and Coordination of Research (FRN), and the Swedish Transport and Communications Research Board (KFB). In addition, the Department of Technology and Social Change at Linköping University made resources available to support the work.

Jane Summerton
Linköping and Berg, Sweden

Foreword

Historians and social scientists, as well as engineers and managers of technology-driven companies, should be particularly interested in and learn from the approaches to technological and social change in this volume. Its contributors see technology in both its multifaceted systemic complexity and its interwoven relationship with other social forces. Managers of research and development and practicing engineers with broad responsibilities for technological change know that a one-dimensional approach to complex technological problems is simplistic and unworkable.

The conference at which the papers in this volume were presented was one of five held internationally since 1988 on the history and sociology of large technological systems. All have been dedicated to the proposition that the invention, development, and use of technological hardware and software involve a seamless interaction with political, economic, and social activities. The authors of these papers, social scientists and historians, have learned from their first- and second-hand contact with engineers and managers that these professionals rarely reach a goal or fulfill a plan by viewing technology simply as machines, chemical processes, electrical and electronic devices, or material structures. Instead, the experienced system builder perceives technology as sociotechnical systems, sometimes requiring a technical, other times a scientific, and often a social solution to the sequence of problems that arises as technology is being developed.

For instance, my present study of the history of military-funded megasystems since World War II shows that system builders who managed the engineers, scientists, and skilled workers designing and constructing the production systems that made weapons found themselves facing nearly insurmountable political and organizational problems as often as they encountered technical ones. Most of the system builders in charge had conventional engineering-school educations that poorly prepared them for the diversity and complexity of presiding over technological change, but they learned to cope with the reality of their responsibilities by using the tacit knowledge and broad perspective that their practical experience provided.

Narrowly trained engineers often mistakenly assume that technological change involves only technical problem solving; narrowly focused social scientists and historians disinterested in technical matters and working without contact with engineers and scientists also often mistakenly assume that technological change can be satisfactorily explained with reference only to its social causes. Fortunately for the reader seeking deeper and broader understanding, the authors in this volume realize that technology shapes society and that society shapes technology as well. A majority of those attending the conference were present at the earlier conferences in the series and have enthusiastically accepted the proposition that the papers presented and conversations conducted must be formulated and developed free from the distorting approaches that present technology as being its own sweet beast or as shaped only by social forces.

The high quality and vitality of the conference organized by Dr. Jane Summerton in Sweden in August 1992 resulted in part from the support given in Sweden to social scientists and historians studying technology. The department at Linköping University to which Dr. Summerton belongs has an internationally renowned science and technology studies program, and the Royal Institute of Technology in Stockholm has a broad-visioned history of technology program headed by Professor Svante Lindqvist. The Royal Swedish Academy of Engineering Sciences honors historians of science and technology with membership and has a committee active in promoting the study of the interactions of science and technology. Nils Eric Svenson of the Bank of Sweden Tercentenary Foundation greatly stimulated the flowering of science studies and the history of technology in Sweden through prudent and imaginative funding. This broad institutional support helps explain why this major international conference on technological systems could take place in Sweden; it also promises that Swedish social scientists and historians will continue to show an international community new and fruitful ways of studying technological and social change.

This volume is the third anthology to emerge from the large technical systems conference series (earlier volumes were edited by Mayntz and Hughes [1988] and LaPorte [1991], see the introductory essay for complete references). Having had the good fortune to participate in all of the conferences, I perceive several intellectually commendable trends. Views about the development of large systems held by senior scholars in the early conferences have regularly been revised by younger scholars taking part in the more recent conferences. For instance, I argued in the initial conferences that large technological systems as they age and increase in size take on momentum and, therefore, strongly resist changes in their character and direction of development. The papers given at the Swedish conference and included in this volume provide ample evidence that a number of forces bring about substantial changes or reconfiguration in large systems.

Another positive development in the study of the history of large systems has been the regularly increasing number of scholars involved. At early meetings, we could list many questions about large systems that needed to be answered, but there were few scholars and even fewer published essays with answers. Now there is a better match between questions and answers, as the number of scholars concentrating in the subject area regularly increases. I suspect that the stimulating essays in this volume will reinforce this trend.

Thomas P. Hughes
University of Pennsylvania

1

Introductory Essay:
The Systems Approach
to Technological Change

Jane Summerton

This book is about processes of change in systems of technology. More specifically, it concerns processes by which the large technical systems of our everyday lives — railways, electricity, telecommunication systems, and others — are reshaped or *reconfigured* in various ways.

At first glance, this phenomenon appears almost counter-intuitive. For most of us, technical systems conjure up images of stability and permanence. In social science research as well, they are often regarded as symbols of the complexity, ubiquity and embodied power of modern technology. Pending system failure or other strong forces, reconfiguration appears unlikely to occur.

Historical and contemporary experience clearly shows, however, that large technical systems can and do undergo processes of reconfiguration. National systems are transformed into transnational ones as they expand over territorial borders. Transnational systems are disintegrated as a result of political developments such as those in the former Soviet Union. Systems or parts of systems with different functions are integrated in new ways with far-reaching impacts, exemplified by the adoption of telegraphy in railroads. In the last few decades, many systems have been deregulated or re-regulated, resulting in some areas in what has been termed the most fundamental industry reorganization since the appearance of large, franchised monopolies earlier in this century.

What factors impact upon large technical systems and induce — or force — such reconfiguration? What can we learn about the dynamics and processes by which transformation takes place (or perhaps is prevented from occurring), actors' responses or strategies, the problems encountered and how these are dealt with? The purpose of this book is to take a step toward answering these questions by analyzing processes of system reconfiguration in different kinds of systems, in different contexts, and over time. The reader will not find a general theory of systemic change in these pages. Instead, the chapters are primarily empirical in nature, drawing upon historical and contemporary case studies. Some of the systems studied are ones we all recognize, others are less known. The ambition is to provide a range of empirical studies that can be used as a base for comparison and generalization.

One point of departure is that an understanding of how and why large technical systems are reshaped is useful for policy makers, managers, engineers, and other *operators* who are responsible for the workings of these systems. There are at least three additional reasons for studying processes of reconfiguration in technical systems. The first is a *user* argument, stemming from the uncontested ubiquity of such systems and our dependence upon them (at least in industrialized societies). Technical systems are integral to our daily lives: they heat and light our homes and offices, fly us to conferences and vacation islands, and allow us to call grandmother on Sundays. In other words, we as consumers are undeniably parts of these systems. When they are reshaped, parts of our lives are reshaped.

The second reason why it is important to understand reconfiguration in technical systems has to do with public policy, political control and our roles as *citizens*. Many technical systems are public systems, if not formally, at least in terms of their service mandates. It is widely recognized, however, that large systems of technology are often challenges to public insight and control; indeed the metaphor of technics-out-of-control is the epitomal example of modern technology.[1] As citizens, most of us harbor a perhaps naive hope that things are not that bad: a measure of democratic insight and some form of regulation is both possible and necessary. Several chapters of this book show, however, that when systems are reshaped, shifts in control or power are likely to occur. It is important to understand the implications of these changes for public policy.

The third motive for studying processes of reconfiguration is a social science *researcher* perspective. It is well known that analyzing extreme situations — system failure, periods of rapid growth or structural transformation — often help us to elucidate "the normal" in different expressions of technology. By studying phases in which technical systems undergo radical change, we might expect to gain new insights into basic dynamics and properties of these systems.

The organization of this essay is as follows. The systems perspective to technology is briefly presented and types of reconfiguration in large technical systems are discussed. Thereafter the various chapters of the book are summarized in relation to the themes they reflect. Following a discussion of driving forces and actors' roles, the essay concludes by suggesting some issues for future research.

Technology as Systems

The point of departure for a systems perspective is that many technologies cannot be viewed as isolated artifacts. Instead they are parts of larger "wholes" (material technologies, organizations, institutional rule systems and structures, and cultural values) which support and sustain them. Seen as an artifact, the telephone is a small box in our kitchens and bedrooms. Viewed as part of a system of cables, exchanges, corporations and global telecommunications links, it is just one visible expression of what might be the largest technical system ever installed.[2] Similarly, to become a car owner is not merely to own an artifact of mobility. It is:

> ...to gain admission to a web of complex sociotechnical systems. To buy a car is, in a real sense, to buy into complex road, energy supply, parts distribution, maintenance, registration, insurance, police and legal systems.[3]

Implicit to the systems approach is thus the perspective that in seamless webs[4] of technology, the "technical" and "social" dimensions of technology are intertwined. The technical is inherently social. In the terminology of this book and elsewhere within the approach, a large technical system carries with it the assumption of the *sociotechnical* character of such technology.

The concepts of "system" and "systemicy" are used in highly divergent ways in the literature, both theoretically and empirically. Beckman (this volume) reviews some of these uses, asking the question of what exactly is "systemic" about technology.[5] As a basis for defining the nature of systemic interdependency in technology, Beckman identifies a number of general types of operations in technical systems. Systems regulation is one approach within social theory for explaining technical change. Beckman argues, however, that there is not necessarily a connection between something being "systemic" and its being systems regulated.

The systems approach to technology has a long history.[6] As Staudenmaier has pointed out, Adam Smith used the machine as a metaphor for the interconnectedness of systems.[7] Marx' philosophy of social systems and the impacts of technology on social relations is well known. Within sociology, important early roots to the view of modern technologies as systems can be found in the works of Ogburn and Mumford. Even before the 1920s, Veblen emphasized the complexity and fragility of the modern industrial system. A systems perspective has also been pervasive within the history of technology for at least the past three decades.[8]

In the mid-1980s, social science and historical interest in technical systems experienced a renaissance. A research approach that focused specifically on *large technical systems* (LTS) emerged. This approach is the intellectual framework for the present book. The single most important work in shaping the perspective was Thomas P. Hughes' pathbreaking *Networks of Power: Electrification in Western Society 1880 — 1930* (1983). Hughes developed a number of concepts that are useful in understanding the dynamics of change in technical systems. One such concept is *system builders*, referring to the inventors, engineers, managers, financiers and others who develop, support and sustain technical systems. System builders use a variety of tactics to promote and defend their systems. They can be expected to block attempts at reconfiguration that threaten their control.

Another concept from Hughes' work that is relevant to understanding system change is *reverse salients*.[9] Reverse salients is a military metaphor for problems or lags that hold back a system on some "front" of its development. A reverse salient, which can be technical, economic, organizational or political in nature, thwarts the further growth of the system. System builders must be able to correctly identify reverse salients and turn them into solvable problems.[10]

Within the past few years, large technical systems have been studied by increasing numbers of researchers in the history and sociology of technology, political science, economics, and other fields.[11] The growth of intellectual interest is also reflected in a series of international round table conferences that have been held among researchers since 1986.

Two contemporary research approaches that are related to the systems perspective in analyzing processes of technological change are actor-network theory[12] and the social construction of technology.[13] Among other things, these approaches emphasize that systems and networks are constructs. They are shaped in contingent and often conflict-filled processes of interpretation, interaction and negotiation among purposeful actors or groups of actors. While much of the large technical systems research tends to treat actors as units *within* the analysis, actor-network theory and the social construction of technology view actors as the explicit units *of* their analysis.

Types of Reconfiguration

This book concerns *reconfiguration* in large technical systems. In some ways this approach is of course not new. As Braun and Joerges point out (this volume), an interest in large-scale and historically unprecedented change has been implicit in most earlier research in the systems field. Similarly, actor-network theory and the social construction of technology are concerned with processes of change by which stabilization or "closure" is achieved in technology.[14]

This book complements these perspectives, taking a somewhat different point of departure. It is well known in the research approaches noted above that periods of stability in technical systems and networks are typically only provisional. Systems and networks are dynamic entities; they can seldom — if ever — be "blackboxed" (closed) for good. The current volume focuses on the dynamics of situations or periods in the development of technical systems in which *previously achieved closure is undone*. This is another way of formulating the classic question of what causes black boxes of technology to open.

The undoing of closure opens up the potential for transformation. The system is in transition from a "business as usual" mode to a new phase in which various aspects of system operation are called into question. In such periods, the inner workings and structure of a system often begin to show. Taken-for-granted assumptions about a system — its function, the configuration of its core technology, its organization, how it is controlled — are perhaps challenged. In the process, competitive battles among actors or even broader social conflicts are likely to occur.

It might be tempting to associate the dynamics of reconfiguration with rapid or revolutionary process. Such an assumption is, however, potentially deceptive. What initially might appear to be a "revolution" can in fact be the outcome of a series of small, incremental adaptations over time. The cumulative effects of these steps can nevertheless be at least as substantial as the effect of abrupt innovation.

From a historical perspective, several types of reconfiguration in large sociotechnical systems can be discerned. One well-known type has been the *territorial expansion and interconnection of similar systems across political borders*, transforming regional systems into national ones and national systems into transnational ones. In this book, the clearest examples of territorial border crossing are the integration of the telecommunications systems of formerly East and West Germany, as well as the interconnection of regional electricity systems in Australia to form a national system.

A second type of reconfiguration has been the result of linkages between systems with heterogeneous functions. Here we see a different form of "border crossing", namely the *crossing of functional system boundaries* by combining parts of different systems that complement each other. For example, transportation systems have been fundamentally changed by linkages with communication systems or energy systems, as witnessed by the use of telegraph systems and electricity in railways or the adoption of radiotelegraphy in ocean transport sytems. Examples in this book are found in the shaping of such disparate phenomena as organ transplant systems and industrial mass warfare systems.

Finally, a third type of reconfiguration in large technical systems can be seen in recent developments in many Western countries, namely the *reorganizing* of previously monopoly systems into new configurations based on principles of competition and open access. This can perhaps be viewed as a form of institutional "border crossing." In this book, the opening of European electricity grids to competition is the clearest example of this type of reconfiguration.

Cultural Embeddedness and the Potential for Change

Like all technology, large technical systems are culturally embedded phenomena. What does this mean and what are the implications for the potential for reconfiguration? The conserving role of cultural values is often emphasized, while the importance of values and expectations in inducing system change is less so. Increased environmental awareness, for instance, is an important factor behind pressures to reconfigure today's road traffic systems, particularly in large cities. Likewise, public perception of risks behind nuclear power have been known to change the direction of energy systems.

One way to approach systems as cultural phenomena and the implications for transformation is to look at expressions of cultural incompatibilities. One form of cultural incompatibility is *incompatibility between the system cultures of different "meeting" systems*. System cultures are ways of thinking and doing things — the traditions, values, and modes of practice of a particular system. Incompatibilities or perceived incompatibilities in the meeting of systems is the topic of several chapters in this book. One case is clashes between national computer networks that were to be interconnected. Another case is the integration of alternative technologies in traditional power systems. In both cases, "technical" compatibilities are in fact cultural ones.

Another form of cultural incompatibility is *incompatibility between a system and its sociocultural milieu*. This dimension is well known from case studies of failed technology transfer from industrialized to third world countries. The clearest example in this book is the systemic failure of full-cycle nuclear power in three industrialized countries.

The various chapters thus analyze processes of transformation in technical systems from a variety of perspectives and themes. Part One centers on the dynamics of crossing boundaries of systems in a functional sense, recombining parts of heterogeneous systems. Part Two concerns reconfiguration by crossing territorial borders. Part Three and Part Four explore cultural dimensions of system change such as cultural incompatibility and other issues. Part Five is the only non-empirical section of the book, exploring the logic of systemic technology from a purely theoretical perspective.

One weakness of any organization of chapters is, of course, that shared themes along one dimension are drawn out while thematic similarities among chapters in different parts of the book tend to be concealed. This should be kept in mind when reading the chapter-by-chapter discussion of themes that follows.

Crossing Boundaries of Systems

Much of the existing literature on large technical systems centers around traditional infrastructure systems in transportation, energy, and telecommunications. The most common topic is the development and growth of individual systems. Relatively little attention has been paid to the ways in which interlinkages between them have influenced how the individual systems have developed.

The chapters in the first part of the book address this neglect. Braun and Joerges point to a previously unexplored aspect of boundary crossing: a case in which interlinkages among heterogeneous parts of systems create an entirely *new* system. Braun and Joerges use the example of the emergence of transborder organ transplant systems. Organ transplant systems are specific, task-oriented large technical systems. They have evolved through a complex networking of "classical" infrastructural systems of road and air transport, long distance data transmission, and telephony. Although organ transplant systems are dependent upon the successful functioning of existing infrastructures, they have independent cultural and institutional identities as systems. Braun and

Joerges propose the concept of "second-order large technical system" to describe this kind of system.

In an application of Braun and Joerges' concept, Bucholz discusses the way in which armies, railroads and information technology were combined to create systems of twentieth-century industrial mass warfare. Armies can be viewed as epitomal examples of large technical systems. The author shows how the use of railroads, supported and integrated by the telegraph and telephone, transformed the traditional European armies in the period leading up to World War I. The result was a second-order military system whose technological and social base "was so huge it dwarfed even the largest civilian systems."

The largest civilian systems today are typically operated by multinational corporations. Schneider suggests that multinational corporations such as IBM and Siemens can be viewed as nested large technical systems within huge global infrastructures. In an argument similar to that of Braun and Joerges, Schneider argues that the global integration and differentiation in today's corporations would have been unthinkable without sophisticated use of modern telephony, air transport systems, satellites, and computer networks. Interlinking and recombining these systems opens up new means of organizing production. The result is that corporate structures are being transformed, becoming more globalized, more tightly-coupled and — perhaps paradoxically from a "classical" systems perspective — more decentralized at the same time.

Linkages among systems or parts of systems are explored from a somewhat different perspective in the chapter by Usselman on signaling in railways. Here the focus is upon the dynamics of systems that are specifically *embedded* in larger systems. Usselman shows how the remaking of signaling in American railways at the turn of this century involved a complex interplay among a hierarchy of different systems. These were the systems of signaling, the individual railroad systems, the national railroad system, and the national regulatory system. Far from being a uniform technology, the embedded system of block signalling was an innovation of considerable variation that "spread in fits and starts" within different railroads in the national system. Its implementation gave rise to many problems and many agendas. Usselman argues that useful concepts in analyzing interactions among systems and subsystems in transition are *indeterminancy* and *differentiation*. The key issue becomes whether changes "fit" into the indeterminancy of the system and whether the differentiation thus caused can be tolerated.

Territorial Border Crossing

The theme of Part Two is reconfigurating by expanding and integrating systems across territorial borders. The integration of the East German and West German telephone systems following the fall of the Berlin wall is a case in point. As Robischon points out, one response to the new situation could have been to allow two separate state companies, as is still the case in railways. System builders defined the goal, however, as one of creating a united and technically homogeneous network throughout Germany. Robischon argues that although state unification provided the political framework, the final design of the system was a direct expression of the motives and strategies of purposeful corporate actors.

Actors' motives and strategies in the politics of change are also a core theme in Salsbury's chapter on current efforts to form a nationwide electricity grid in Australia. To do this, the borders between existing state systems must be crossed. Salsbury argues that the goal of creating a national grid has little to do with improving system load by e.g. matching regions with different characteristics. Instead it is a political solution to severe economic and social problems that have continued to plague both managers and politicians. Salsbury observes that the motives behind Australia's restructuring of its power grid therefore differ greatly from those which spurred earlier "border crossing" in expanding electricity systems as described by Hughes (1983).

The restructuring of electricity grids is on political agendas in parts of the United States and Europe as well. Traditional monopolies are being reconfigured to open the systems to competition. Coutard discusses the myriad of economic isues that are entailed in carrying out these reforms. New principles of equal access call for the "opening" of territorial borders between systems. Coutard points out that integrating grids also requires *disintegrating* institutions and practices (cross-subsidies, reciprocity agreements) among cooperating utilities in different territories. Grid services and the prices paid for them must be "unbundled," raising complex network issues for regulators. Reconfiguring the systems reveals the mechanisms of interdependence among formally independent but cooperating systems.

In some ways, the chapters in Part One and Part Two form a seamless web of argument about the dynamics of reconfiguration through border crossing. Systems *simultaneously* cross territorial borders and the borders of functionally different systems. When intergrating systems such as electricity or telecommunications across territorial borders, it is often necessary to "disintegrate" previous structures. Political integration and functional disintegration go hand in hand, challenging the traditional view in which these trends are assumed to be parallel. In all of these processes, reconfiguration often means *organizational* and *institutional* border crossing as well.

Cultural Dimensions of System Change

Part Three and Part Four center around cultural dimensions of system change. The authors explore cultural expressions of systems in different contexts, analyzing the significance for the way in which the systems have developed. As noted earlier, an important unifying theme in the chapters is the idea of cultural incompatibilities.

Incompatibilities between the cultures of systems can be particularly evident in battles among competitive systems. The interconnection of computer networks in the United States in the 1970s and 1980s provides a vivid example, as described by Abbate. The challenge of defining acceptable technical standards to connect the systems was a clash between two competing systems with highly different system cultures. Abbate argues that to understand this struggle and its impact upon how the systems ultimately were connected, it is necessary to look beyond issues of "technical" incompatibility. Instead the author reconstructs the cultural reference points of the conflicting actors, exploring the goals, expectations and world views behind their arguments and actions. The drawn-out process of negotiation revealed highly divergent perceptions and assumptions among system builders about the social and cultural implications of different technical configurations. "Technical" incompatibilities had deep roots that reflected profound differences in system cultures.

This theme is echoed in von Meier's study of the potential for reconfiguration of traditional electricity systems. Whereas Lovins once argued that "soft" and "hard" energy paths were inherently incompatible, von Meier argues that today's advanced "supple" technologies (wind power, photovoltaics, solar thermal systems) are neither technically nor economically incompatible with existing power systems. Instead, they can provide clear benefits, offering sophisticated means of increasing flexibility and relieving a range of system constraints. Von Meier observes that actual use of these technologies in power systems to date has, however, been impeded by a range of cultural biases, including *perceptions* about incompatibility on the part of engineers and managers. As von Meier notes, "to the extent that these beliefs are not justified by factual evidence, they...constitute a type of cultural incompatibility."

Cultural incompatibility can also be expressed as incompatibility between a system and its sociocultural milieu, as discussed earlier. This dimension is addressed by Rochlin in a cross-national study of the implementation of nuclear power in France, Germany, Sweden and the United States. Rochlin argues that the full-cycle nuclear system — as originally designed in these countries — was a sociotechnically inflexible system that placed extraordinary demands upon conformity and long-term stability in the societies it was intended to serve. It was only in France that crucial sociocultural conditions existed and remained stable enough over time for nuclear power to be successfully deployed as planned. In the other three countries, lack of compatibility between the intended system and its milieu — in combination with the inability of the system to adapt — ultimately led to what Rochlin calls "artifactual success but system failure."

Nuclear power is an extreme example of a tightly-coupled large technical system. In contrast, Grundmann and Juhlin each explore the cultural embeddedness of a system that is decidedly more loosely coupled, namely the road traffic system. The road traffic system has been largely neglected in studies of technical systems.[15] On one level, the two chapters by Grundmann and Juhlin respectively are highly similar. Each author examines "futuristic" new technologies for road management that are (perhaps) on the verge of being introduced, asking what impacts that these technologies are likely to have. The chapters differ considerably, however, in their theoretical perspectives and the level of the system upon which they focus.

Grundmann analyzes the potential for reconfiguration from the perspective of "would-be" system builders within the automobile industry and their patterns of conservative innovative activity to date. In his analysis, Grundmann draws upon conceptual tools from three current schools of thought on systems and networks. These are the large technical system approach (Hughes), the actor-network approach (Callon, Latour), and the interorganizational network approach (Joerges). The author argues that although visions are global (focusing on problems of pollution, congestion etc), the proposed solutions will depend on local cultures of engineering and the ability to create networks and political alliances. The outcome is difficult to predict. The car traffic system is set apart from other large systems specifically by *the extent of its cultural embeddedness*. Grundmann concludes, "since the car traffic system probably mobilizes more passions than any other LTS, the conflicts that emerge around its restructuring will be profound and long lasting."

In contrast to Grundmann's global or industry-wide approach, Juhlin analyzes the road transport system as a system based on local interaction. To understand road transport as an operational system, Juhlin argues, it is necessary to leave the macro-analysis of most system approaches. Instead the researcher must go to the micro-level — the perspective of the individual driver. Juhlin explores traffic as symbolic interaction, asking how a variety of new road transport technologies might impact upon traffic culture and the autonomous and competitive social framework of driving. Drawing upon Douglas' typology of social organization, Juhlin proposes two ways in which reconfiguration of the system might occur, concluding that the new technologies may be able to discipline today's aggressive drivers into a "common traffic culture."

The chapters by Grundmann and Juhlin thus have very different points of departure. Grundmann takes a "top-down,"automobile industry perspective while Juhlin chooses a "bottom-up," individual driver one. The authors nevertheless agree that the road traffic system has been extremely loosely coupled to date, lacking a system builder with the capacity to coordinate it. They also agree that the proposed new technologies for road traffic are more than technological solutions to problems in today's road system. They are mechanisms for changing *traffic culture*. Reshaping the system will entail changing deeply embedded attitudes and patterns of social behavior. Today's driver is king of the road and not likely to give up this role.

Driving Forces?

A common concern in virtually all of the chapters in the book is to understand causes of system change. It is clear that large technical systems reflect varying types of dynamics in different phases of their development. Some phases are characterized by steady growth and relatively conflict-free development, during which the system is successively taken for granted. In phases of reconfiguration, however, a system which had previously appeared "invulnerable" undergoes dramatic change. What opens up the black box?

A detailed discussion that would do justice to the authors' sophisticated handling of complex interplays among factors and events in various systems is not possible here. Instead the following is meant to serve as a brief indication of aspects that are interwoven in the various chapters.

It is clear that factors or events which on one level might appear to "cause" change — for example, the introduction of a new technology — are often an expression of underlying problems within the system (reverse salients in Hughes' terminology) and actors' views of these problems. Reverse salients can be viewed as, on the one hand, real or "objective" problems in the form of technical inefficiencies, rising costs, limits in capacity, or problems of reliability. On the other hand, as MacKenzie[16] and others have pointed out, reverse salients are far from given. Actors typically have different perceptions of where the system is going or should be going, where the problems lie, and how these problems can and should be solved. The ways in which problems or reverse salients are identified, defined and resolved reflect interests and interpretations. They also indicate the ability of various actors to carry out or impose their own agendas of change.

These perspectives are found in varying extents behind many processes of reconfiguration, as indicated in the case studies of changes in computer networking, railroad signalling, and electricity grids. As well, the studies of the introduction of new information technology in road traffic systems, as well as the "technological comeback" of wind power in power systems, also show that actors perceive problems and solutions in rather different ways.

One problem, or reverse salient, that is particularly striking seems to be *congestion* in the physical network of many systems. Acute problems in *system flow* on railroad tracks, computer networks, or roads need to be solved. Sometimes the task has been viewed as one of leveling the load. In other cases system builders have strived to increase the capacity of the network. In still other systems (such as car traffic systems and armies), the sheer complexity of movement among huge numbers of interacting units has meant finding means to organize flow. Here the solution was not necessarily to increase network capacity. Instead it has been to develop mechanisms for increasing the regularity and predictability of movements.

Reverse salients (or perceived ones) can also arise from what economists refer to as negative externalities. Externalities of large technical systems include their *environmental impacts* and risks.[17] As noted earlier, environmental concerns are salient factors behind emergent processes of change in road traffic systems and electricity systems. Similarly, concerns about safety have reshaped systems with perceived high accident rates in the past, such as early railroads. Public criticism of externalities are thus often expressed as *consumer pressures* to alleviate the problem.

Because of the public service character of many large technical systems, consumer demands predictably lead to *political pressures* to improve performance or make the system cleaner, more accessible or more affordable. These pressures sometimes result in regulatory measures that transform the system, as in contemporary cases of deregulation. Regulators grapple with the enormous complexities involved, while engineers and managers seek strategies to handle the unsolicited regulation. Broad *political ideologies* are often the backdrop for this type of system reconfiguration. Systems such as public infrastructures become instruments for carrying out political agendas.

Likewise, *political developments or contingencies* in an even broader sense (such as war or the threat of war) can also radically alter the shape and direction of large technical systems. Perhaps the clearest examples in this book are the transformation of armies prior to World War I and the integration of East-West German telecommunications.

Finally, it is well known that *changing competitive conditions* among systems can force reconfiguration. System builders must develop strategies to defend and expand their markets (as in the case of multinational corporate systems), find other markets (as witnessed historically in the competition between railways, cars, and airlines) or become obsolete (such as town gas in many areas).

Actors' Roles:
Patterns in Changing Webs

The preceding section touched upon actors' roles in various ways. What does this book indicate about actors' agendas, alliances, and strategies in processes of reconfiguration?

A logical starting point is to ask who are the system builders in the reshaping of systems. The chapters in this book indicate very different answers in various types of systems.

It is obvious that transborder systems elicit the emergence of various kinds of *supranational* actors. Supranationals can be huge networks of formally sovereign but functionally interdependent organizations. They can be corporate actors in tightly-coupled, centrally controlled global firms. They can be transnational regulatory institutions such as the EC or other regulators with new "border crossing" mandates. Examples of the latter are found in such diverse processes as the remaking of American railroad signalling and the restructuring of the Australian electric power industry.

Big systems are not necessarily operated by big actors, however. Some large technical systems are neither centrally coordinated nor tightly coupled. Juhlin argues in this volume that an understanding of the operational system of road traffic requires a shift in perspective from the type of large organizations usually studied in large technical system research to *individuals* as operators — in this case drivers. This is made vividly apparent when considering the prospects for reconfiguration. Any attempt to transform the system must take into account the deeply-embedded attitudes and behavior of the individual drivers. It must also consider the social framework in which they interact. Thus individuals as operators — rather than as consumers or citizens — are a second important category of actors in reshaping systems. What are the methodological implications of this largely-unexplored dimension in systems studies? Can a focus on the individual and the social framework be a means, as Juhlin proposes, of studying how the social integration of individual operators takes place even in highly centralized and tightly-coupled large technical systems?

From a different perspective, systems such as road traffic suggest a third answer to the question of who are potential system builders. In road traffic, there appears to be no system builder in sight who has the capacity to "unify all the recalcitrant elements" that would be needed to successfully introduce the electric car. Instead it has been suggested that what coordinates the road traffic system is the technical logic of the system itself. All the participants buy into the same logic — or expressed another way, the same understanding — of the system, knowing that the other participants also do so and carry out their roles accordingly. In this kind of system, change would appear to depend upon building *networks or alliances* among manufacturers, consumer organizations, traffic controllers and other actors with shared interests. Thus we might expect constellations of heterogeneous actors between the "top-down" (supranational) and "bottom-up" (individual operators) levels, where no one has systemwide control.

Indeed, the politics of reconfiguration in large technical systems often elicit unexpected alliances and ideological mixes. A Labor party and a Conservative party unite in the restructuring of an electricity system because it is a strategy for fulfilling a shared, broader goal. In railroad signalling, the means of solving one problem in railways (congestion) became the ends for those looking to solve another problem (safety). One change can sometimes suit many actors' agendas, thereby uniting actors and groups who perhaps have conflicting goals in other ways. Compatibility of interests is stronger than their conflicts, forming the basis for coalition formation. This is one of the ways in which the transformation of a system can lead to new dependencies among groups of actors.

At other times, a reform that on the surface appears to offer benefits to all actors is blocked by latent conflicts that are not immediately apparent. This theme was reflected in several chapters. Reshaping technical systems can be expected to produce conflicts on at least two levels. The first is conflicts between actors — between system builders of competitive systems, or managers and regulators, or managers and consumer groups — as they struggle to protect and promote their interests. The second is "generalized social conflict."[18] Examples are seen in polarized debates over the deployment of nuclear power, as well as predictions of similar debates about new road traffic systems.

The potential for societal conflict touches on the seldom-explored emotive dimension of certain large technical systems. They have a potential for mobilizing our passions, to use Grundmann's words. Reshaping (or attempting to reshape) a system can be explosive, triggering deep rooted and long-lasting social conflicts.

When technical systems undergo change, issues of control are often actualized. Reconfiguration can lead to shifts in the distribution of power. Amidst the flux, system builders can be expected to develop strategies and mechanisms to retain or increase their *managerial control*. Corporate leaders in multinational firms, for example, make sophisticated use of telecommunication networks as global technologies for managerial control. An older strategy of system builders in striving to assure managerial control is to limit the number of compatible or "acceptable" technical options. Certain actors are thus included in reconfiguring the system, while others are excluded. Here the rhetoric of defining the new system (what is possible, what will work) can be as important as actual system design. Finally, in a strategy we recognize from earlier system development, system builders often "go along with" limited legislation or regulation in order to ward off more extensive regulatory control. One gives up control of parts of the system to preserve control of the whole. Each of these strategies appears in this book.

Strategies on the part of system builders to increase managerial control thus raise the question of how to insure democratic insight and a measure of *sociopolitical control* on the part of the societies served by the systems. Nowhere is this challenge expressed more clearly than in the chapter by Rochlin in explaining the success of nuclear power in France and its failure in Germany and the United States. Rochlin states that where the three countries differed was

...in the degree to which the centralized, "proministrative" technocratic system was open to public participation; in other words, the outcome depended upon the ability or inability of the technocratic elite to keep the system closed off from more general social and political influences. (p. 249, this volume)

The message is clear: for operators, blackboxed systems are more manageable than "open" ones. This type of strategy might be deceptively shortsighted, however; a system without open channels for public feedback is vulnerable to conflict in the long term.

Where Do We Go from Here?

This book offers an array of case studies of processes of reconfiguration in large technical systems. In order to make generalizations about why certain systems evolve as they do, we need to pay attention to differences and similarities in the dynamics of these systems and what they mean. From an emphasis on empirical studies, the systems approach thus needs to further develop and refine its conceptual tools.

In addition, at least three neglected issues can be proposed as priorities for future research. The first is the idea of *stagnation*. Stagnation as a phase in evolving systems has been unexplored within the systems approach.[19] And although stagnation can be an underlying factor behind reshaping systems, few of the studies in this book explicitly discuss this theme. Some systems clearly experience periods of non-growth or decline, however, while similar systems in other areas do not. One example is railroad systems, which in some areas have expanded and in other areas almost "died out," another is nuclear power. What causes stagnation, and what explains the apparent ability of some systems to recover while others fail? Here the potential value of comparative research[20] is apparent. By studying patterns of stagnation and change in similar types of systems in different societal contexts, we should be able to find clues to their causes and implications.

A second underdeveloped theme is the *role of users* in inducing or forcing system transformation. Users are conspicuously absent in studies of large technical systems,[21] somehow unnoticed among the managers, engineers, and regulators. There are many examples of system builders' attempts to shape (or reshape) user behavior or expectations, but can we find cases in which users — through their practices and demands — have explicity reshaped systems? Perhaps traditional "heroic" system builders will be replaced by environmentally motivated groups of users (or other "antiheroes") as the builders of tomorrow's reconfigured systems. Finally, in viewing systems from the perspective of the user, we need to differentiate among heterogeneous users and their (differential) access to power. In this regard, gender perspectives — yet another neglected theme in system studies — would be highly useful, alongside traditional factors such as class, status, age.

A final area that begs explication is *implications for public policy* of research results. How can "lessons learned" be used to influence the development of systems in directions that are desirable from the standpoint of the citizens, users and enlightened operators mentioned at the beginning of this essay? Phases of system instability are opportunities for policy input. Promoting the public interest might mean giving new configurations a chance by supporting weak actors who offer promising alternatives but lack the power to implement them.[22] It might also imply actively promoting measures to increase the *redundancy* of a system as a safeguard against system failure. A high level of redundancy strengthens the robustness of a system. Redundancy in a technical sense is not enough, however; it can be argued that a crucial task for policy is, as well, to develop means of assuring organizational redundancy within the system.

The case studies of this book have another implication as well. The study of "classic infrastructures" has until recently been the study of the evolution of such systems into aggregated, centralized and hierarchial entities. Public policy making, as well, has traditionally been predicated on a regulatory approach based on this view. Studies of reconfiguration show that there is no longer any particular direction or structure towards which or into which old systems will evolve or new ones develop.[23] This is perhaps the most profound policy implication of reconfiguring systems.

Acknowledgements

I thank Svante Beckman, Boel Berner, Lars Ingelstam, Bernward Joerges, Arne Kaijser, Donald MacKenzie, Göran B. Nilsson, Jennifer Nelson and Gene Rochlin for useful comments on an earlier version of this essay.

Notes

1. See for example the writings of Mumford, Ellul and Winner.
2. Joerges 1988, p. 21.
3. Hannay and McGinn 1980, p. 28.
4. See Hughes 1986.
5. See also Joerges 1988 for a discussion of concepts and issues within the systems approach.
6. A comprehensive overview of the literature on technical systems is beyond the scope of this essay. Examples from different disciplines include Trist et.al. 1963, Ellul 1964, Chandler 1977, Rosenberg 1976 and 1982, Hughes 1983 and 1987, Perrow 1984, Joerges 1988, LaPorte 1988, MacKenzie 1990, Mayntz 1993, and Gras 1993. For useful overviews, see Hughes 1980, 1983 pp. 7-9 and Westrum 1991 pp. 73-83.
7. Staudenmeier 1985, pp. 69-71.
8. See Staudenmaier 1985.
9. See Hughes 1983, 1987 and MacKenzie 1987.

10. Law has referred to this type of multifaceted problem-solving in technical networks and systems as "heterogeneous engineering". See Law 1987, 1988.

11. See Bijker, Hughes and Pinch (eds.) 1987, Mayntz and Hughes (eds.) 1988, and La Porte (ed.) 1991.

12. Main texts include Callon 1986, Callon 1987, Law 1987, and Latour 1988.

13. For a few examples among many, see Bijker and Pinch 1987, Mack 1990, and Misa 1992. For overviews of the three approaches to technology, see Bijker, Hughes and Pinch (eds.) 1987, Westrum 1991, Bijker and Law (eds.) 1992, and Summerton 1992.

14. See Bijker and Law (eds.) 1992.

15. For a study of the electric vehicle in France from an actor-network perspective, see Callon 1987.

16. MacKenzie 1987, pp. 196-199.

17. See for example Perrow 1984 and Weingart 1991.

18. Joerges 1988, p. 26.

19. I thank the participants of the Vadstena conference, and in particular Arne Kaijser and Olle Edqvist, for this point.

20. Here Hughes' work remains the model of comparative work within the systems approach with its treatment of the development of the same system (electricity) in different societal contexts.

21. Exceptions in this volume are the chapters by Abbate and Juhlin.

22. See Kaijser, Mogren and Steen 1991.

23. I thank Gene Rochlin for this point.

References

Bijker, Wiebe E. and John Law (eds.). 1992. *Shaping Technology/Building Society: Studies in Sociotechnical Change.* Cambridge, Mass: MIT Press.

Bijker, Wiebe E., Thomas P. Hughes, and Trevor Pinch (eds.). 1987. *The Social Construction of Technological Systems.* Cambridge, Mass: MIT Press.

Braun, Ingo and Joerges, Bernward (forthcoming 1994). *Technik ohne Grenzen.* Berlin: Suhrkamp.

Callon, Michel. 1986a. "The Sociology of an Actor-Network: The Case of the Electric Vehicle." In *Mapping the Dynamics of Science and Technology: Sociology of Science in the Real World,* eds. M. Callon, J. Law and A. Rip. London: MacMillan Press, 19-34.

_____. 1986b. "Some Elements of a Sociology of Translation: Domestication of the Scallops and the Fishermen of St. Brieuc Bay." In *Power, Action and Belief: A New Sociology of Knowledge?* , ed. J. Law. London: Routledge and Kegan Paul. 196-230.

_____. 1987. "Society in the Making: The Study of Technology as a Tool for Sociological Analysis." In *The Social Construction of Technological Systems,* eds. W.E. Bijker, T.P. Hughes, T. Pinch. Cambridge, Mass: MIT Press, 83-103.

Chandler, Alfred D. Jr. 1977. *The Visible Hand: The Managerial Revolution in American Business*. Cambridge, Mass: Harvard University Press.

Ellul, Jacques. 1964. *The Technological Society*. New York: Knopf Press.

Gras, Alain. 1993. *Grandeur et dépendance: sociologie des macrosystèmes techniques*. Pars: Presses Universitaires de France.

Hannay, N. Bruce and Robert E. McGinn. 1980. "The Anatomy of Modern Technology." *Daedalus*. 109 (1): 25-53.

Hughes, Thomas P. 1980. "The Order of the Technological World." *History of Technology*, 1-16.

_____. 1983. *Networks of Power: Electrification in Western Society 1880 — 1930*. Baltimore: Johns Hopkins University Press.

_____. 1986. "The Seamless Web: Technology, Science, Etcetera, Etcetera." *Social Studies of Science*, 16: 281-92.

_____. 1987. "The Evolution of Large Technical Systems." In *The Social Construction of Technological Systems*, eds. W.E. Bijker, T.P. Hughes, T. Pinch. Cambridge, Mass: MIT Press, 51-82.

Joerges, Bernward. 1988. "Large Technical Systems: Concepts and Issues". In *The Development of Large Technical Systems*, eds. R. Mayntz and T.P. Hughes. Frankfurt: Campus Verlag; and Boulder: Westview Press, 9-36.

Kaijser, Arne; Arne Mogren and Peter Steen. 1991. *Changing Direction: Energy Policy and New Technology*. Stockholm: National Energy Administration (now NUTEK).

Kaijser, Arne. 1993. "A Research Approach for Understanding the Development of Infrastructural Systems". *Flux*. 11 (January-March): 53-54.

La Porte, Todd R. (ed.) 1991. *Social Responses to Large Technical Systems: Control or Anticipation*. Dordrecht: Kluwer Academic Publishers.

_____. 1988. "The United States Air Traffic System: Increasing Reliability in the Midst of Rapid Growth". In *The Development of Large Technical Systems*. eds. R. Mayntz and T.P. Hughes. Frankfurt: Campus Verlag; and Boulder: Westview Press, 215-244.

Latour, Bruno. 1987. *Science in Action: How to Follow Scientists and Engineers through Society*. Cambridge, Mass: Harvard University Press.

Law, John. 1987. "Technology and Heterogeneous Engineering: The Case of Portuguese Expansion". In *The Social Construction of Technological Systems*. eds. W.E. Bijker, T.P. Hughes, T. Pinch. Cambridge, Mass: MIT Press, 111-134.

_____. 1988. "The Anatomy of a Socio-technical Struggle: The Design of the TSR2". In *Technology and Social Process*, ed. B. Elliott. Edinburgh: Edinburgh University Press.

Mack, Pamela. 1990. *Viewing the Earth: The Social Construction of the Landsat Satellite System*. Cambridge, Mass: MIT Press.

MacKenzie, Donald. 1987. "Missile Accuracy: A Case Study in the Social Processes of Technological Change." In *The Social Construction of Technological Systems*., eds. W.E. Bijker, T.P. Hughes, T. Pinch. Cambridge, Mass: MIT Press, 195-222.

_____. 1990. *Inventing Accuracy: A Historical Sociology of Nuclear Missile Guidance.* Cambridge, Mass: MIT Press.

Mayntz, Renate. 1993. "Grosse Technische Systeme und Ihre Gesellschafts-theoretische Bedeutung." *Kölner Zeitschrift für Soziologie und Sozial psychologie,* 1.

Mayntz, Renate and Hughes, Thomas P. (eds.) 1988. *The Development of Large Technical Systems.* Frankfurt: Campus Verlag; and Boulder: Westview Press.

Misa, Thomas J. 1992. "Controversy and Closure in Technological Change: Constructing Steel". In *Shaping Technology/Building Society: Studies in Sociotechnical Change.* eds. W.E. Bijker and J. Law. Cambridge, Mass: MIT Press, 109-139.

Mumford, Lewis. 1934. *Technics and Civilization.* New York, Harcourt and Brace, 1934.

Perrow, Charles. 1984. *Normal Accidents: Living with High-Risk Technologies.* New York: Basic Books.

Pinch, Trevor J. and Wiebe E. Bijker. 1987. "The Social Construction of Facts and Artifacts: Or How the Sociology of Science and the Sociology of Technology Might Benefit Each Other". In *The Social Construction of Technological Systems,* eds. W.E. Bijker, T.P. Hughes, T. Pinch. Cambridge, Mass: MIT Press, 17-50.

Rosenberg, Nathan. 1976. *Perspectives on Technology.* London: Cambridge University Press.

_____. 1982. *Inside the Black Box.* Cambridge: Cambridge University Press.

Staudenmaier, John M. 1985. *Technology's Storytellers: Reweaving the Human Fabric.* Cambridge, Mass: MIT Press.

Summerton, Jane. 1992. *District Heating Comes to Town: The Social Shaping of an Energy System.* Linköping, Sweden: Linköping Studies in Arts and Sciences.

Trist, E. L. et. al. 1963 {1987}. *Organizational Choice: Capabilities of Groups at the Coal Face under Changing Technologies.* London: Tavistock Publications.

Weingart, Peter. 1991. "Large Technical Systems, Real-Life Experiments, and the Legitimation Trap of Technology Assessment: The Contribution of Science and Technology to Constituting Risk Perception." In *Social Responses to Large Technical Systems: Control or Anticipation.* ed. T. Rl LaPorte. Dordrecht. Kluwer Academic Publishers.

Westrum, Ron. 1991. *Technologies and Society: The Shaping of People and Things.* Belmont, California: Wadsworth Publishing Company.

Winner, Langdon. 1977. *Autonomous Technology: Technics-out-of Control as a Theme in Political Thought.* Cambridge, Mass: MIT Press.

Combining Parts of Systems

2

How to Recombine Large Technical Systems: The Case of European Organ Transplantation

Ingo Braun and Bernward Joerges

Crossing Borders

On a late night in April 1993, an airplane carrying a donated liver on its way from Birmingham to Edinburgh crashed.[1] Liver, pilot and co-pilot were saved from the bottom of the Firth of Forth by divers of the British Royal Navy and the precious part was duly delivered to be implanted into a 25-year-old woman.

Such stories begin to suggest how organ transplantation involves the transgression of borders. We will discuss technologies that make possible the crossing of considerable spatial and temporal borders required by organ transplantation on the scale practiced at present and presumably largely surpassed in the future. But the building up and dynamics of European and other *trans-border organ transplant systems* (TOTS) can hardly be understood without taking into account the crossing of borders — and attendant blurring of categories — between body and machine, gift and commodity, individual and collective ownership, moral duty and abomination, death and life.

Many of the public problems of transborder organ transplantation arise outside the medical sphere at the far end of extended systems. Two examples are news about "organ tourism"[2] on the first official demarche of a surgeon to operate a commercial organ bank and reports of "organ hunting" in Honduras.[3] Quite possibly, these are instances of the "urban myths" that have sprung up around organ transplantation, circling the news media around the world and being retold like modern fairy tales. But there are more solid fringe phenoma. An example is the organization of transplantation enclaves in pre-war Kuwait where a local legal vacuum, American surgeons, West German medical apparatus, Indian hearts, and a little money combined in the production of organ transplantations which could hardly have been achieved within regular health systems. At the same time, the United States debates whether the transfer of American organs to countries allowing organ implantation but not organ explantation (such as certain Moslem countries) should not be controlled and restricted.

We render such stories at the outset to create a sense of the extent of the system we are going to discuss and to dramatize the turn towards technology. Focusing on the role large-scale technical systems play in setting up TOTS means to systematically decenter typical controversies and to draw attention to the machineries on which TOTS are based. We will show that these machineries are in good part non-medical large scale technologies, on which today´s organ transplantation parasites. When we first presented TOTS at the *Wissensschaftszentrum Berlin*, it was not readily understood that TOTS could be a case in point for the study of large technical systems since "organ transplantation (was) surely at the stage of manufacture technologically," as one discussant put it. On the contrary, we argued, organ transplantation rests (as yet uniquely within the health domain) on a particular type of nested large technical network.

Understanding the Growth of Large Technical Systems

This volume on large technical systems' dynamics is concerned with transformation. Of course, more or less radical transformations have been a prominent theme in large technical systems analyses from the beginning (Mayntz/Hughes 1988, La Porte 1991). In a way, the need to understand large-scale and historically unprecedented transformations has been at the very root of most research in the field. In the process, numerous mechanisms contributing to these transformations have been proposed or identified[4] and it seems to us that mainly three heuristic strategies have been put to work: internalist, externalist, and co-evolutionary. Ours will be a variant of the co-evolutionary perspective.

Internalist strategies assume auto-catalytic mechanisms within systems. One looks for system properties for making increases of scale and diversification of services. Thus, dispositions and control over resources by dominant actors, economic pressures in load management, technological reverse salients, organizational inertia, and similar mechanisms have served as explanatory devices for system transformation.

In contrast, *externalist strategies* highlight the functional dependence of large technical systems on their ecological and societal environments. To explain growth and transformation, one looks for positive feedback mechanisms linking the emergence of networked large-scale systems to the development of all kinds of smaller-scale technologies both in production and consumption. Especially various non-technical environments — economic, political, scientific, cultural and so forth — are used to control large technical system growth and transformation.

Of course, most analyses combine internalist and externalist viewpoints. Seldom, however, has conceptual attention been drawn to large technical systems as an important environment for large technical systems, although historical accounts abound with examples. Let us call strategies looking for mechanisms of transformations at the level of interrelationships between functionally different large technical systems co-evolutionary. Several co-evolutionary growth dynamics may be identified almost a priori. In the first instance, competition between technologically different systems serving similar functions comes to mind. The case of rival gas and electricity systems at the turn of the twentieth century is a well-known example. Also, linkages and complementarities between large technical systems serving closely related functions are familiar enough. The symbiotic growth of air traffic and telecommunication, railroads and telegraph, or railroads and electricity have been variously described. Sometimes complementarity turns into full technological integration or fusion of initially functionally different large technical systems, for instance ISDN in telecommunications.

Recombining heterogeneous parts of large technical systems into new systems with their own internal or external institutional identity is another co-evolutionary variant. We present TOTS as a case in point for such multiple border crossings between several large technical systems. Recombined systems of this sort supersede classical *first-order large technical systems* and constitute what we will call *second-order large technical systems*. First-order large technical systems refer to the familiar, relatively easily delimited all-purpose infrastructures such as the road, railroad, energy, and telecommunication systems that have been at the center of large technical systems research. By contrast, the concept of second-order large technical systems refers to the process of networking parts of different first-order systems for specific, macro-level social domains (see also Braun 1993, Joerges 1992). Much of today's large technical systems' expansion and transformation can be interpreted as superimposing second-order large technical systems on more or less stabilized classic infrastructural systems. We will return to a more abstract elaboration of this concept at the end of the chapter.

Our Approach to Studying TOTS

The transborder scale of organ transplant technology requires and produces a host of locally problematic "harmonizing" requirements across participating health systems. This should include a progressive and internationally uniform metrication of the human body, a legal and technical normalization of organ explantation, a definition and diagnosis of death, and standardized forms of costing. Will the technical, legal and economic normalization of TOTS be accompanied by a levelling of local cultural meanings and symbolizations? What changes will occur in relevant professional systems and control over health policies? What will be the role of local health systems and to what extent will a technically integrated European TOTS be centralized or dispersed politically? How will the difficult issues of unequal organ distribution and non-medical selection criteria be resolved throughout TOTS?

Such questions lurk in the background of our study. While we will provide some tentative answers, our main interest is to show to what extent the underlying processes are conditioned by the growth of a large technical system of organ transplantation — conditioned in the sense that they owe the system their very existence. The ethical, moral, legal, economic and professional problems associated with so-called brain death, for instance, result from TOTS technology since brain death was invented in TOTS as a major means to increase and speed up organ supply.[5]

We will concentrate on non-medical network technology and describe the processes of technical extension TOTS went through. We will not spell out in detail non-technical (such as organizational, economic, legal, and cultural) loops since we are emphasizing technical system change. Of course, ours is not an engineering or medical discourse either. Rather what we try to do is to mark the technical elements generating non-technical issues and non-technical issues generating technical solutions. We also set out the promises of new technical solutions and the resulting technical problems justifying new technical advances. In short, we will point out cross-overs and re-entries between technical and non-technical spheres of TOTS without tracing the entire spiral.

Brain death may again serve as an illustration of this appoach. If our study concentrated on the local medical technology of organ transfer (which it does not), then a detailed description of technical change in death diagnosis would be offered. We would show how the prevailing legal definitions of death became impractical and were supplanted by the notion of brain death. We would describe various technical definitions of brain death and the technical steps that allowed for a progressive shortening of the time span required for verifying the diagnosis and we would point out how this enabled TOTS to increase the supply of transplantable hearts. We would also mention the required legal changes, ethical debates about the legitimacy of brain death procedures, and the resistance of the nursing profession to take prolonged care of brain-dead patients. Similarly we would discuss — hypothetically now — the new push for artificial heart research resulting from mounting public criticism of brain death practice.

However, we would not look into the processes by which brain death and the transplantation of human organs became culturally legitimate practice or how — again hypothetically — its broad acceptance began to erode. In other words, we would not retrace in great detail important parts of the movement from initial awe and horror (as pictured in the early movies featuring organ transplantations) to general acceptance plus irritation with non-professional and criminal practice (as reflected in the movies and TV serials of the 1980s). While we would deal at length with the technical problems of looking into the brain, we would only indicate the process by which Christian churches came to officially support brain death based organ transplantation, telling believers that transplantation does not interfere with resurrection and that offering the gift of an organ is a caring act. We would only mark the effect on the technological agenda of debates linking brain death issues to abortion issues and transplantation to genetic engineering, debates in which churches oppose TOTS and may therefore help to decrease organ supply, setting in motion further searches for technical ways out.

TOTS Technology

TOTS were built up starting in the late 1970s and are partially based on air and other fast transport systems, telecommunication and computer network systems, and various high-tech surgical and chemical clinical technologies. A chronic undersupply of donor organs has driven the past development of the transplant system. This has caused a series of turbulences in the system. At present, European TOTS confront a range of problematic issues which have to do with their transnational integration and clearly necessitate further transformations with only some foreseen with any confidence.

The very notion of second-order technical systems implies that the technical substrate of TOTS is heterogeneous and that system borders are fuzzy. Any attempt at classifying and circumscribing must be highly provisional. We are motivated by the wish to set out the basic transorganizational network character of TOTS and not to contribute to a description of a much lamented *Apparatemedizin*.

TOTS as Technically Based Inter-Organizational Networks

In the past, health technology studies have largely been micro in that they focused either on behavioral implications for medical personnel and patients or on various health organizations and firms. Attention has been drawn to the technical linkage of clinics and to a lesser extent praxes with all kinds of medical apparatus and equipment. Interorganizational networking of these technologies is in full progress. Inter- or supra-organizational networking (as described by industrial sociologists for the machineries operated by producer, supplier and sales companies) is being introduced in some countries. For the time being, the phenomenon remains restricted to parts of the health system that deal with finance and insurance and the regionally organized logistics of medical emergency services.

In clinical medicine, TOTS represent a technical development where for the first time interorganizational and transregional technical networks play an important role (see Schoeppe 1989: 27ff.). Thus, we will not speak of the diffusion of transplantation technology (a notion that might be plausible in cases like the diffusion of computer tomography) but about the large technical system of organ transplantation and its extension.[6]

The technical networks of transplantation link a wide spectrum of varied technologies — transplantation specific ones and other medical and non-medical technologies. The networks link reliable, routine technologies with experimental ones. They link highly heterogeneous mechanical, chemical and biological technologies and put them to work in a host of organizations, such as transplantation centers and hospitals, tissue and organ banks, dialysis clinics and laboratories, transplantation coordination and information centers, and the many medical emergency services within the transport, airline and telecommunications networks. This linking-up of technologies and their organizations is achieved mostly through existing communication and transport infrastructures.

The resulting transplant organization is spatially and functionally widely dispersed. It requires a higher level of normalization as compared to local hospital organizations in the standardization and compatibility of the apparatus utilized across the entire organization, data formats, drugs, transport containers, conservation devices, surgery and treatment, and routine cooperative exchanges between the compo-

nent organizations and professions involved. Compatibility of bureaucratic formats also serves to legitimatize the central control and monitoring of the hospitals involved.

Compatibility of bureaucratic formats is in turn highly technical,
for instance in the application of standard tests for tissue typification
and the exchange of computerized data. PIONEER, the computer communication system introduced by the Eurotransplant organization in
the late 1980s to link non-medical and medical transplant organizations in the participating countries (see Broom 1988), plays a key role
in these standardization processes. PIONEER links into Eurotransplant's central mainframe in Leiden thirty-eight transplantation centers and forty-one typification laboratories in five Eurotransplant
member countries (Germany, Austria, Belgium, the Netherlands and
Luxembourg). The system stores over 100,000 immunological patient
profiles, data of potential transplantation recipients (state of health,
degree of urgency and treating hospital), as well as data about results
of particular transplantation therapies. Whenever a donor organ becomes available, PIONEER selects a suitable recipient and cross-
checks regional waiting lists and Eurotransplant's comprehensive lists
of possible patients.

The operational characteristics of the technically networked trans-
plantational field constrain local organizational autonomy. The pattern is not unlike some military modes of operation. There are long periods of normal, relatively quiet and entirely routine clinical activity
that alternate with short phases of hectic activity where clinical routines and organizational hierarchies are subverted (see Rochlin, La
Porte, Roberts 1991 and Rochlin 1989).

During slow periods, the network is used to find potential donors and
donor organs and ascertain certain post-operative and evaluative functions. As soon as it has identified a suitable donor, a veritable cascade
of synchronous activities is released. This includes procurement of consent for transplantation, preparation of explantation and implantation, information exchange regarding donor state and organ specification, the matching of donor tissue information to potential recipients'
tissue information in the data pool, computerized selection, physiolog-
ical compatibility tests, and explantation. Finally test tissues, organs,
patients, doctors and apparatus must be transported and the organ implanted. To the local organizations, these activities seem like sudden
attacks on the normal day-to-day clinical process and require the mobilization of all kinds of resources during a very short period of time.

Conflicts often arise from extensive communication needs. Supply organizations, coordinators and explantation teams are all anxious to
keep in touch with their transplantation center, thereby blocking the

telephone system of hospitals that might be ready to offer help. Plenty of examples illustrate these problems. In some cases they have provoked efforts to formalize cooperation among hospitals.

Systems such as PIONEER operate their networking functions through existing infrastructure systems for long-distance data transmission. Transplantation operations are, therefore, heavily dependent on highly efficient networks. But TOTS also depend on closely-meshed transportation networks, especially road and air. If the transplant or the donor's body, the explantation team, the tissue sample and the recipient have to travel only a short distance, it is common to use ambulances or taxis. But if the donor is far away from the hospital, other forms of transportation are needed. About one-third of organ explantations make intensive use of air transport, especially for multi-organ explants in which various organs of a donor are brought to separate regional transplantation centers and several transplantation teams from different centers are involved. As a rule, media coverage of transplantation celebrates this technical aspect. Indeed there is no other field of medicine where everyday work routine is so much determined by the extensive use of airplanes and helicopters. Similar to hospitals specializing in emergency medicine, a landing field for helicopters is mandatory at organ transplantation centers. In the U.S., where greater distances have to be travelled for organ transfers, some transplantation centers even have their own small airports.

Apart from modern long distance data-transmission networks and air-traffic systems, the good old telephone is used extensively. The telephone is virtually essential during the busy phases of transplantation, when explantation procedures have to be coordinated with organ transfer and implantation procedures. When due to a software error, the AT&T network broke down in early 1990, transplantation activities largely came to a halt in the U.S. Participating hospitals could no longer be coordinated and it became impossible to locate relatives of deceased patients to obtain their approval for donating organs.

TOTS as a Technically Based Interorganic Network

As the acronym TOTS implies, one can see transplanting itself as a networking of bodies and organs. In the process of transplantation a technical network is established between several human bodies. These are usually the body of a deceased person and several bodies of recipients. As we shall see later, this act of networking holds the critical point for all popular ideas about the questions of life and death. The bodies to be linked with each other are determined through a so-called tissue matching procedure, a trial body coupling.

The central technical object within this body networking is the organ to be transplanted. The transplant's *technical* character is mostly overlooked in public debates about transplantation medicine. Instead, emphasis is on the morally connotated categories of natural versus artificial organ substitutes (*Organersatz*). But donor organs are themselves protheses, artificial limbs. As soon as the dead donor body is kept functioning through an intensive-care machinery, one must speak of these organs as technical products. The intensive care treatment of potential donors means frequent and differentiated control of lab test data to monitor metabolic and circulation functions. The organism lacking cerebral functions is controlled by apparatus and medication designed to keep up the functions of the organs to later be transplanted. In addition, transplanted organs require continuous technical care and maintenance and life-long follow-up examinations, immune suppression medication or immune modulation (see Pichlmayr 1986).

Beyond this, networking of bodies is also taking place on another level. In participating organizations, donor and recipient bodies are connected to technical systems and their transorganizational networks. How far-reaching the connection of professional and everyday handling of bodies is and how far this technical network reaches into the clients' everyday lives is vividly illustrated by the beeper system.

When potential recipients — mostly kidney transplant patients — have moved close to the top of a waiting list they receive a beeper. As long as they carry this gadget, they can be reached at any time and any place in case a suitable organ becomes available; in the meantime, they can move around freely. Once they have been called on the beeper, they must immediately get in touch with their local dialysis center and call a taxi to the transplanting hospital or possibly the nearest airport. Members of transplantation teams in turn are called on their emergency beepers and moved to the locus of the transplantation.

The networking of bodies extends to *potential* donor bodies and that means almost everyone. Apart from supporting the transplantion operation and the integration of donors' organs in recipients' bodies, the transplantation system also procures donor organs. A considerable section of TOTS technical networks, therefore, serves to provide organs. Or rather it serves to connect to the system any human body within its reach about to face death.

The technical linkage of donors is based mostly on virtual capacities of the network, thus augmenting the chances for the recovery of organs. This fact, however, does not help to relieve peoples' fear and bewilderment when they are confronted with the transplantation system. It is not so much the fear of possibly being a transplantation candidate or knowing that transplantations are performed as the thought that through "virtual networking" everybody has been turned into a potential donor.

TOTS' Medical and Fringe Technologies

A complete study of TOTS technology would require at least two more sections dealing with its *medical* apparatus and technical fringes, namely where it becomes difficult to determine system relations.

As to the first, TOTS not only combines various technical infrastructures but also an extensive, heterogeneous range of medical apparatus, instruments, machines and pharmaceutical products in explantation, organ transfer and implantation. These include the following:

1. instruments for speeding up and increasing the reliability of brain death diagnosis;
2. extensive intensive-care machinery for artificial respiration, automatic monitoring of circulation and for maintaining brain-dead donors' metabolic functions and circulation until the organs are ready to be explanted;
3. so-called transplantation survival systems now under development to ease time dependency;
4. many technical apparatuses for the typification of donors' and recipients' tissue samples;
5. artificial organs, such as artificial hearts, artificial kidneys and pulmonary machines;
6. appliances and procedures after implantation for preventing rejection of transplants through drugs such as the famous Cyclosporin and the newly developed FK 506.

An interesting aspect here is that system growth feeds on system growth, namely that chemical intervention to preserve organs ruins other organs and calls for the organ's replacement in due time.

The *technical fringes* of TOTS — for example the radio or phones used to coordinate transplantation activities — show how interwoven systems such as TOTS are with technical installations of modern societies. Ordinarily one would not describe the radio or traffic as being part of this system. As a technical framework, however, they do sustain and shape the system's operations. One more example may suffice. Victims of accidents, particularly traffic accidents, are the largest group of donors. In particular, motor biking is considered a "net-organ supplier" meaning that bikers belong to a group of people defined by age and way of life and death who supply more organs than are needed by this same group. The reason for the classification is the healthy age of the victims and the chance that most organs remain uninjured during the accident. People dying in car accidents are often locked in wrecked cars and suffer burns and organ lacerations. Motorbikers are flung into the air and suffer mainly head injuries. As the saying goes, "motor cycles are donor cycles" since brains are not yet in high demand for transplantation.

There is another way large-scale technology links into TOTS. Disastrous accidents such as mass collisions on highways, airplane crashes or the Chernobyl nuclear accident strain national and international transplantation systems. Large technical systems both cause disasters and help handle them.

Transborder Issues in TOTS

According to large technical system history, classical technical infrastructures have been built up in a rather erratic though clearly expansive, bottom-up way. Historical texts as well as ongoing social narrative usually use an evolutionary rhetoric in describing their growth. In this view, the starting points of a system's build-up are different local centers for demonstration and experimentation. Step-by-step, these are linked up to regional networks, which in turn are integrated into national and international infrastructural systems.

Together with an increase in regional scale, the organizations controlling the networks (and state interest in these organizations) grow. Underlying technical change is conceived as a scaling-up process with familiar smaller scale technologies being enlarged and adapted to the operational requirements of massively networked structures.

Top-Down Expansion

The story of TOTS is a little different. The first organ transplantations were performed in the 1970s at top national health systems in well funded research projects at the best clinics. The medical technology used was experimental, which is still the case in many areas of transplantation medicine.

Since effective immunosuppressive drugs were not yet available, early organ transplantation consisted mostly of kidney transfers between close relatives. Occasionally other organs from non-relatives or dead donors which were sufficiently compatible were transplanted. Regional and international organizations for targeting suitable transplants had not yet been established. The donor pool was limited to the area covered by a particular hospital where the research project was located. Transplant exchange took place mostly on the level of international research groups. The few transplantations occurring with organs not coming from relatives resulted from such international exchanges. Coordination of traffic and communication between participating organizations — nowadays a formalized routine — was more or less improvised.

The formation of the international Eurotransplant Foundation in 1967 exemplifies how transplantation medicine took off internationally. During the first years, its main tasks were to initiate research in all member countries in transplantation-related medicine (such as vascular surgery, histology and immunology), coordinate on-going clinical research projects, and assist with the supply of donor organs. By the 1970s, most of the surgical difficulties had been overcome, tissue matching had been mastered, and the creeping destruction of transplants after implantation could be met by suitable drugs. It was time to establish a technical-organizational network on national and later regional levels. To be sure, the physicians involved did not have to start from scratch. Existing reference systems provided organizational means of networking. There were facilities for kidney analysis, blood donation and emergency-rescue services, which were well developed and widespread in most industrialized countries.

Early on, transplantation surgery was shaped by a few outstanding phycicians. The first to transplant a kidney in the late 1960s was Joseph Murray, who laid the foundations for tissue-matching and immunosuppression. Liver and lungs were first transplanted by Thomas Starzl and James Hardy in 1963 and Christian Barnard pioneered transplanting a heart in 1967, an event that triggered an unprecedented coverage in the media. These physicians, however, do not really match the notion of "system builder" put forth by Thomas P. Hughes. Dedicated doctors, non-profit organizations and private foundations joined in pushing ahead the building of the technical network of today's transplantation practice. It is interesting to note that non-medical organizations and particularly associations of potential users (associations of heart and kidney patients and their relatives) have greatly contributed to set up the system.

Thus, the growth of TOTS has little in common with the history of classic infrastructure systems. It resembles more the process of expanding infrastructure systems by adding special subsystems, such as the addition of certain facilities that combined roads and rail goods traffic to the existing railway system. Another example is the expansion of an existing broadcasting system by adding facilities for long-distance data communication to an existing telephone system. In their study, Mayntz and Schneider point out that the introduction of video text in France, England and Germany did not spread from local islands but was centrally mapped out, planned, controlled and systematically pushed forward.[7]

System Differentiation: Denationalizing or Nationalizing?

In the final analysis, incongruities and disparities between organ supply and demand are due to a certain biological fact that human bodies are highly incompatible. This fundamental tissue incompatibility is the very basis of international organ transfer. For if the immunological fitting of donor and recipient tissues were not relevant for transplantation, national or sub-national systems could only assure sufficient transplant supply.

This suggests that the present task of TOTS consists in the compensation — or at least the attenuation — on an international scale of regional and national tissue incompatibilities and in solving the resulting allocation problems. At the same time, specific system dynamics in part threaten its international character and vary its action radius. The instability is mostly due to spatially and temporally dissimilar development of contributing sub-systems and technologies. The number of transplants procured internationally decreases as national and regional systems are developed in countries where transplantation surgery originated. The requirement of international organ exchange is also decreasing through the results of immunological research. Immunosuppressives such as cyclosporine help in attenuating the compatibility requirements that must be met by donor and recipient tissues (see Braun et. al. 1991). Finally, the international system is increasingly by-passed by experiments with animal transplants, so-called autotransplantation and human biological cryotechnologies. The development of tissue-culture transplants is beginning to benefit from the latest achievements in genetic engineering research.

Legal assimilation of different national definitions of life and death and organ donation regulations of TOTS act in similar ways. In the Netherlands, for instance, the explicit consent given by donors during their lifetime is a prerequisite for organ explantation, which is a condition that impedes organ procurement. In neighboring Belgium, organs can be recovered in principle from all dead persons who have not during their lifetime explicitly opposed explantation. As a consequence, Belgium supplies more organs to the Dutch than vice versa. When such national differences are leveled out, the extent to which donor organs have to be transferred between different countries is accordingly decreased.

Counteracting such nationalizing effects, trends favoring further expansion of supranational TOTS can be observed. Increasing routinization and simplification of transplantation surgery result in more and more countries taking up this branch of medicine. Frequently, this happens before an adequately dimensioned donor system is built up. In fact, the establishment of a comprehensive, culturally acceptable and politically safeguarded donor system requires a considerably longer preparatory period than the build-up of transplantation clinics solely used for implantation. As a consequence, the transplantation medicine of newcomer countries remains in its initial stage dependent on international organ supply systems.

More sophisticated transplantation surgery goes hand-in-hand with differentiation processes in organ demand. With regard to more popular donor organs such as kidneys and to a certain extent hearts and livers, it is not only the availability of some transplant that matters but more and more that of *high-quality* transplants. Above all this means obtaining organs from donors who are as young as possible. Donor-organ quality requirements that are not specifically dependent on immunological necessities seem to become more exacting, thus narrowing the logistic margin opened by the lowering of biological tissue-compatibility requirements and creating new incompatibilities premised on non-medical criteria.

Technical achievements in immunomodulation and transplantation surgery permit the gradual extension of the range of routinely transplantable organs. As a consequence, for each newly admitted organ — such as spleen, brain tissue, pancreas, lungs, testes, ovaries and bone marrow — the initially required transnationality could be restored. Eventually, the most important stimulus for maintaining international system integration may be a specific change in function affecting transplantation systems since the late 1980s. The logistics for recipient selection were established in the 1970s on the basis of the medically and logistically necessary large-area tissue compatibility imperative then. With more effective drugs for immunomodulation, logistics began to serve monitoring functions, increasingly performing activities of verification and legitimation. More and more, they now support establishing economical and other non-medical selection criteria, facilitating keeping undue competition between transplantation centers under control, and monitoring transplantation and organ assignment practices that differ widely between countries and clinics.

In our view, the solution to the countervailing interests and require-ments of operational range will be a functional differentiation within TOTS. Internationally integrated system levels will probably be ori-ented to more general, less organ-specific tasks than previously. They will essentially be limited to the monitoring functions described above and to operating the exchange of exotic donor organs for experimental purposes, namely organs that are seldom transplanted and whose sup-ply systems with a national radius would be insufficient or too costly, and for transplantations considered for research rather than routine therapeutic purposes. This may include organs needed for acutely life-threatened special patients and innovative transplantations that must meet extraordinary requirements in terms of tissue compatibility or legitimation.

For the transfer of less exotic organs and transplantations, rela-tively stable and organ-specific systems will be established nation-ally and even locally. The situation can be compared to the modal-split problems in traffic and communication systems undergoing a pro-cess of differentiation such as the distribution of traffic flows to indi-vidual and public transport systems or of TV transmissions to cable and satellite systems.

TOTS as a Second-Order Large Technical System

Keeping close to our illustration, we will now examine structural dif-ferences between transplantation systems (and other second-order sys-tems) and associated transport and telecommunication systems (first-order systems) on which the former are based. We will then discuss in more general terms certain repercussions on the development of first-order systems such as transport and telecommunication systems from the progressive build-up of second-order systems such as TOTS.

Interdependent Technical Heterogeneity

TOTS represent large scale technical systems similar to classical in-frastructural systems for the following reasons:

1. TOTS' assigned functions are executed mainly through pre-existing technical networks spanning large spatial and temporal distances;
2. their operation is a precondition for the functioning of a large number of small technical systems;

3. many socially important operations are carried out via their networks;
4. they link many different, otherwise largely unconnected, organizations and actors.

At the same time, however, the example of TOTS suggests a series of differences because:

1. Transplantation networks heterogeneously recombine other networks;
2. classical infrastructures are geared to a multitude of different uses, while organ transplantation systems are limited to very specific tasks;
3. TOTS and similar systems can claim only a little technical or organizational "proper substance."

The latter point means that their technical networking and supra-regional operative scope depend on services rendered by classical infrastructures, in particular the telephone, data communication, road transport, and civil and military aviation. TOTS networks are largely borrowed from other more extensive technical systems. The proper contribution of TOTS to networking consists then of *combining* existing infrastructural facilities. In this process, TOTS operators use each infrastructural system as it would be used by any household or business but they also intervene in the operation of the linked systems. For instance, they receive preferential treatment with airlines for the transport of medical personnel and patients, operate rental jet planes, and enjoy privileged access to emergency services or to various telecommunication facilities.

In short, first-order and second-order systems are similar in the scale of technical operations carried out through these systems and the spatial range of their frequently international operations. They differ in that second-order systems are set up for specific purposes and possess relatively little technical/organizational proper substance or scope.

First-order systems are relatively well insulated from malfunctions in other linked systems. When the telephone system breaks down, travelers can still use railroads even though it may be more difficult to plan further activities at their destination point. If railroad personnel go on strike, people can still use cars even though in such a situation serious traffic congestions can be expected. Second-order systems, on the other hand, are more prone to functional failure since they depend on the simultaneous functioning of classical infrastructural systems. Organ transplantations can be delayed or prevented by the collapse of just one of the infrastructural elements involved.

By linking first-order systems, each system's performance levels and inherent operational problems are partly combined and accumulated. For this reason, second-order systems are generally more vulnerable and need relatively higher degrees of standardization. For this reason, second-order systems tend to change more rapidly than first-order systems.

Less proper technical substance of second-order systems also contributes to their relatively rapid establishment and change. Second-order systems are temporally more flexible than first-order systems. The West German TOTS were drawn up in a little less than a decade — a relatively short time span when compared to the time required for the installation of nationwide power or water supply systems. Nor will it be easy to find in classical infrastructures a parallel to the rapid functional change of TOTS in the second half of the 1980s.

Co-Evolutionary Effects

How are classical infrastructures affected by the progressive superimposition of second-order systems? As second-order systems utilize existing infrastructures in new ways, it is almost self-evident that their build-up goes hand-in-hand with growth prospects and higher performance requirements for each of the combined infrastructural systems. Thus, in joyful anticipation of massive expansion, in the second half of the 1980s the German Telekom solicited the interest of certain companies and industry associations by explicitly referring to the new possibilities of long-distance networking for telecommunication services provided by ISDN. It even asked potential customers to come up with their own imaginative solutions. (The financial burdens of German unification considerably hampered these attempts.)

On the other hand, traffic planners are concerned about the increased requirements on performance that result from big industrial enterprises introducing just-in-time production regimes and shifting product stocks and associated costs on to roads already heavily loaded with freight traffic. In Germany, it is now argued that freight carriers should pay more for road system extension and maintenance.

Second-order systems partly affect first-order system operation and thereby potentially stimulate infrastructure innovation and flexibility. Developed and financed largely by private-company associations, new German systems of freight-train utilization and wagon monitoring are currently being established. TOTS can also serve as an example of how the effects of first-order system innovation triggered second-order system development. In the 1980s, this phenomenon pioneered the European-wide use of paging devices and long-distance data communication.

Finally, as second-order systems combine the operation and utilization of different types of infrastructures, they may facilitate convergence on infrastructural developmental levels. This speculation assumes that internal mechanisms repairing Hughes' "reverse salients" — considered a central feature of classical infrastructural systems — are also operating in second-order systems.

Technical Homogeneity, Zero-Order System, Interconnections

In general, the concept of second-order systems is meaningful only when the underlying notion of the large technical system is maintained within certain limits of specification. Speaking of the large technical transport or energy supply system, for example, is much too unspecific since the margin for potential combinations is too narrow. On the other hand, speaking of the large technical bus shuttle system of Linköping Airport in Sweden or the large technical electric power grid for industrial consumers at Vadstena is too specific with the margin for combinations being too wide. The notion of second-order systems is potentially useful when the large technical system concept is at an *intermediate* level, permitting for example distinctions between the large technical systems of spatially extended railways and road traffic, gas and power, and water and sewage utilities.

In distinguishing between first-order and second-order systems we apply *technical homogeneity* as a major criterion next to specificity of purpose. Tacitly, large technical systems' literature has assumed that normally these systems can be treated as technically homogeneous. In other words, they are distinguished from each other on the basis of characteristic unmistakable core technologies. With some systems it is comparatively easy to agree on the typical network technology. For railways, it is rails and trains; for road transport, roads and automobiles; and for power grids, cables and electricity. But even for systems whose technical network structure is less massive, uniform or apparent (such as in air traffic or non-cabled telecommunication systems), this approach to system classification raises difficult problems. Solutions derived from engineering literature fail because the nexus between technical and non-technical aspects of systems is almost systematically effaced.

Providing such problems of empirical delimitation can be solved, the concept of first-order and second-order systems can be supported and extended by *radicalizing the criterion of homogeneity*. First, following Ekardt's suggestion (1994) and not trying to look for different technologies typical of different large technical systems, one could seek the one technology common to all large technical systems. Indeed, with *structural engineering and the construction industry* we have a type of infrastructure that the large technical systems debate, despite its predilection with construction and system building, has hitherto almost totally neglected. To emphasize the complementarity of such large technical system substructures in relation to second-order system superstructures, Ekardt has named them *zero-order large technical systems*.

These zero-order large technical systems above — or more aptly below — all account for the ever deeper penetration of technical systems into their ecological base. By the same token, the conceptual excommunication of this zero-order social level helps in keeping consistent a currently fashionable theory that an era of heavy technological burdens on the environment is now turning into one of environmentally more benign technical dematerialization, heralded mostly by innovations made possible by telecommunication systems. For example, Sherman and Judkin (1992:151) conclude from their studies of information technologies that "(c)ompact, highly cabled electronic offices will replace the large corporate office blocks which will stand only as monuments to previous business philosophies....Not a happy thought for the construction industry." Having discovered the "virtual workplace," Pruitt and Barett (1991) find that it will only be situated where the workman is situated, namely in "cyberspace," which is both global and fits into a tiny electronic box.[8]

The next step toward a more general explication of the second-order systems concept sets out more explicitly the *interconnection* phenomena central to these systems and places them within the endless spectrum of potential interconnections of seamless webs. Note here that second-order systems as often as first-order large technical systems refer to (extra-somatically based) technical interconnections as opposed to non-technical ones such as economic, organizational and other institutional links. As with non-technical links, there are different types of technical interconnections, depending mainly on whether they connect technological cores or peripheries of large technical systems. The first case is the coupling of single nodes (or node connections) of two or more utility networks with one of the involved systems playing the role of the user and the other that of the performance supplier (see Figure 1). Correspondingly, one of them establishes and operates the interconnection. For example, in railway network operation, power grid connections are made through special transformer stations and, conversely, power grid fuel supply is insured through special rail links. Such point-specific system fusions can be found in all large technical systems. They normally do not affect the utilization of the interconnected systems in any significant degree and in this sense one might say that all large technical systems are second-order in the first place.

System I   **System II**

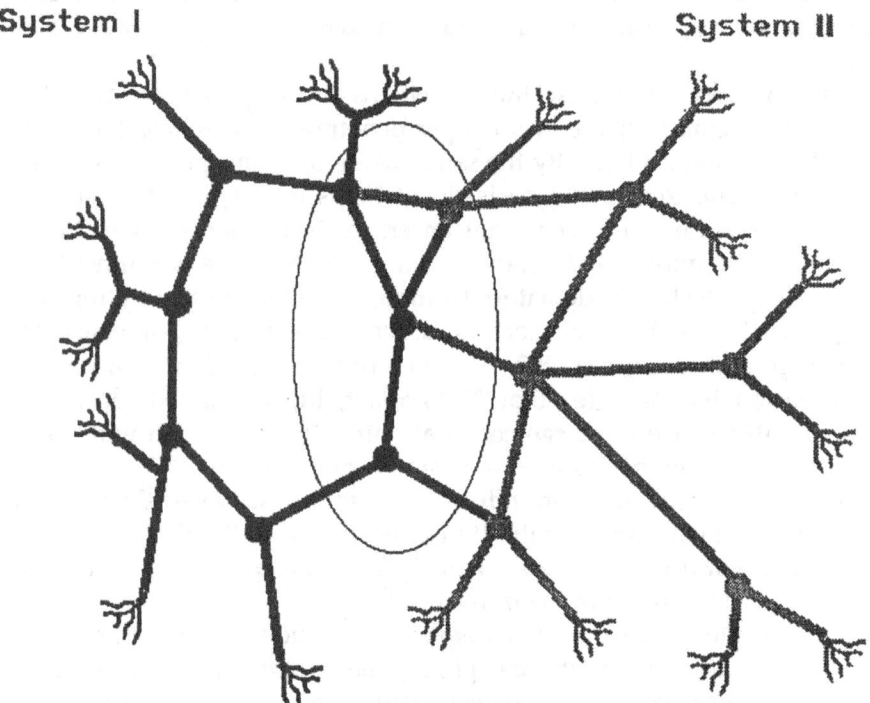

FIGURE 1 Interconnections between network cores.

The other case is the technical coupling of loose network ends (see Figure 2). It is not really a case of interconnecting large technical systems but of some user organizations establishing and operating interconnections. Firms and households establish links between the infrastructure networks accessible at production sites and homes. This will allow interlinking power, water and sewage grids for the operation of such household items as washing machines. Such point-specific interconnections are a prerequisite for the operation of most small technical systems and they normally do not significantly affect the operational mode of the large technical systems involved.

System I   **System II**

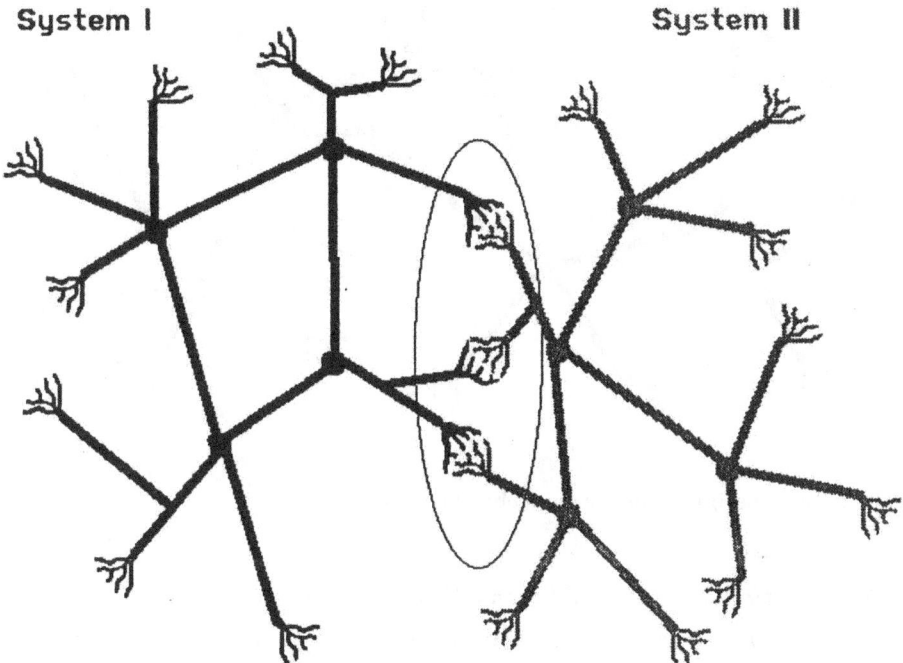

FIGURE 2 Interconnections between network peripheries.

Second-order systems can be *positioned halfway between these two poles* (see Figure 3). Loose network ends and the cores are technically linked with the organizational competence for establishing and operating the interconnection lying outside the large technical systems. Thus, in technical terms, they are situated between the interconnected large technical systems. In organizational terms, they are situated between large technical system operators and users. The concept of second order large technical system is meant to emphasize this *double hybrid* character.

System I   System II

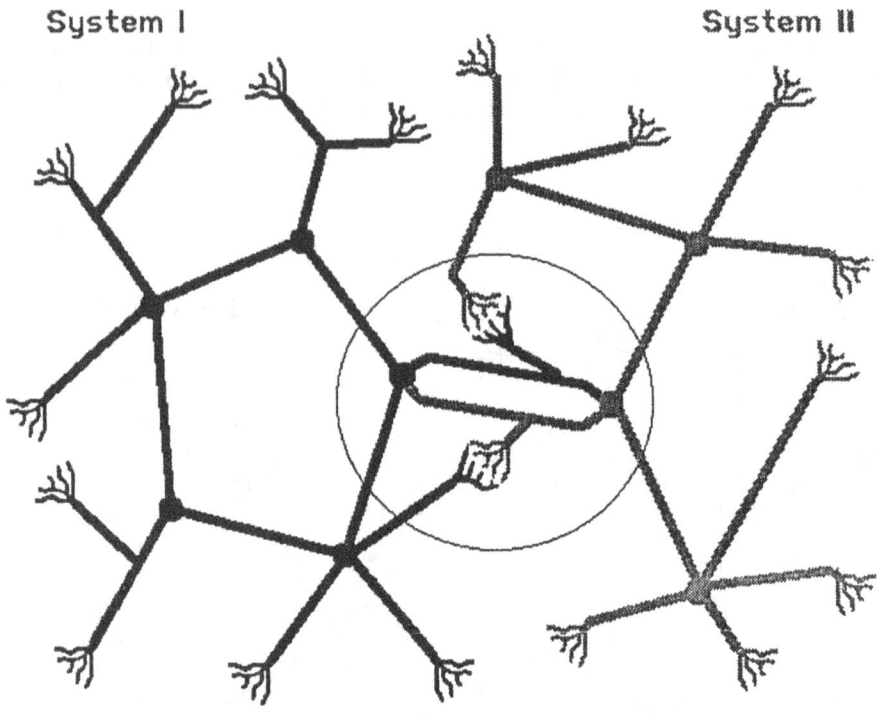

FIGURE 3 Second-order large technical systems.

Co-Evolution of First- and Second-Order Systems

The utility of conceptualizing a particular type of large technical system in terms of its heterogeneous technical linkages remains to be seen. It requires analyzing its behavior over time and showing that the development of different types of large technical systems is interrelated in a way that validates these distinctions. Does the distinction help us understand the evolution of large technical systems insofar as the development of first-order and second-order systems are interrelated in theoretically interesting ways? We will briefly point to three possible answers that are compatible with the TOTS case and the concept of second-order large technical systems. We will call the three versions the contingency, precursor and differentiation theses.

The *contingency thesis* is the most conservative because it maintains that first-order and second-order systems develop independently of each other. In this view, first-order systems are an evolutionary precondition of second-order systems or else second-order systems are evolutionary by-products of first-order systems. Potential co-evolutionary effects are limited to the above-mentioned growth, innovation and convergence dynamic. The strong claim of this thesis lies mainly in the assumption that second-order systems are on a par with other large technical systems, namely that they can be regarded in their own right as a large technical system type. Accordingly, major conceptual problems have to do with delimiting and comparing both system types.

The *precursor thesis* views second-order systems as independent in only a restricted sense. In accordance with the general rule that new technology often develops out of old technology, second-order systems are understood here as precursors for new large technical systems. New first-order systems can — but must not necessarily — develop out of them. According to this thesis, second-order systems with time tend to become less purpose-specific and technically more homogeneous. Thus, in the end, the number of first-order system species increases in an incestuous way. In support of the precursor thesis, the genesis of some classical infrastructure systems can be cited. For instance, the first railway lines were still purpose-specific facilities for bridging gaps in transportation, i.e. for establishing links between existing road and waterway networks. One conceptual challenge here lies in demarcating small and large technical systems. Does it make sense to distinguish second-order systems from extended configurations of small-scale technical systems? Which non-incestuous developmental paths lead from smaller to larger technical systems?

The *differentiation thesis* — the most radical of the three propositions — views second-order systems not as characteristic of the earlier stages of systems development, but rather as indicative of a new phase in the history of large technical systems. In accordance with another rule of thumb, namely that new technology is a model and precondition for the further development of old technology, the thesis assumes that first-order systems tend to become more purpose-specific and technically more heterogeneous over time and metamorphose into second-order systems. Thus, in the end the few purpose-unspecific large technical systems would be replaced by a frayed tissue of multiple large technical system phenomena only distinguishable by their specific purposes. Basically, this assumes that the organizational demonopolization, dispersal and deregulation typical of the late growth of large technical systems will be followed by technical heterogeneity. This would not necessarily lead to new organizational orders (smaller, decentralized units) but to a "new loss of orientation" (*neue Unübersichtlichkeit*[9]) in the large technical systems landscape. This can be seen in the more recent modernization of large technical systems such as air traffic and telecommunication (for the case of German telecommunication, see Kubicek 1994). The challenge of this line of reasoning comes from the difficulty of distinguishing large technical systems from each other and from smaller systems irrespective of the technologies in question, i.e. solely with reference to system goals and functions and the organizational processes involved.

Conclusion

Organ transplantation shares with many other late industrialist societal subsystems an essential dependence on technical and non-technical large-scale, networked operations. We have given two meanings to the term large scale. One concerns the aspect of a highly dispersed, non-local (or transborder) temporal and spatial reach; the other concerns the systemic aspect of interdependent technical heterogeneity. This second aspect was highlighted as constituting a particular type of technical system which we called second-order large technical system.

The build-up of the European organ transplantation system is a story of continued expansive moves meant to solve chronic structural problems resulting in great part from persistent organ scarcity and time pressures. We have drawn attention to a great variety of moves, each promising a solution to specific technical problems induced by earlier moves. We have indicated where technical reconfigurations produced cross-overs into non-technical problem areas and where non-technical moves produced re-entries into technical layers of TOTS. The system has extended quickly in the past fifteen years or so. The up-scaling of its technical network of networks was largely parasitic on existing technical infrastructures and involved ever more effective immunosuppression. It was so successful in terms of its interested parties that at the time of this study its future course is uncertain. The technical networking of organizations, organisms and organs will certainly continue but might be put to different uses of monitoring, verification and legitimation.

The consecutive technical moves to mobilize more organs and gain time in transplanting them — always justified by the need to close the gap between organ supply and demand — have in fact led to a widening of the gap. Year after year more and more kinds of organs are procured, but year after year more people with more kinds of indications come into the system. TOTS is a growing large technical system of the second order that is driven by technical solutions to problems arising from technical advances and by the particular cultural accommodation of this dynamic. At each stage it reproduces on a larger scale the problem that has set it in motion. There is little reason to expect that this basic dynamic of TOTS will lose momentum in the foreseeable future.

In telling the story of TOTS, we have used many terms and rhetorical devices generated by those agencies who manage or otherwise legitimately speak for the systems under study. While this is almost inevitable in studies primarily based on written documents, it should be noted that the practice produces a considerable "systemic bias." The kind and degree of "systemicy" (Beckman, this volume) described in our study is generated within certain layers of the system and more or less successfully imposed on others. In the world perspective of the crashed motor cyclists "donating" their hearts, the system presumably appears less well structured. Generally, we must assume that seen from the underside, the seamless web (Hughes 1986) shows more loose ends and blurring of patterns.

Notes

1. *Frankfurter Allgemeine Zeitung*, No 80, 5 April 1993.
2. *Washington Post*, 19 March 1983.
3. Die Tageszeitung, Berlin, 12 July 1988.
4. For an overview see Joerges/Braun (1994).
5. For details see Braun et al. (1991).
6. For a general critique of the diffusion concept in organization change research see Czarniawska-Joerges (1993).
7. Mayntz/Schneider, however, tend to consider videotext (BTX) as a autonomous large technical system beside the telephone and other systems, not just an extension or addition to the existing telephone system (see Mayntz/Schneider 1988).
8. For more of this see Sotto (1993).
9. A term famously introduced by Habermas to suggest the latest condition of modernity, implying either that systemicy is lost or that it escapes us.

References

Braun, I. 1993. "Geflügelte Saurier. Systeme zweiter Ordnung: ein Verflechtungsphänomen großer technischer Systeme." In *Technik ohne Grenzen*, ed. I. Braun and B. Joerges (forthcoming).
Braun, I., Joerges, B. 1990. "Körper-Technik: Zur Wiederkehr des Körpers durch technische Erweiterung." In *Argumente, Sonderband 182*, 83-104.
Braun, I., Feuerstein, G., von Grote-Janz, C. 1991. "Organ-Technik: Technik und Wissenschaft im Organtransplantationswesen." In *Soziale Welt 4*, 445-472.

Broom, G. 1988. "Further information on PIONEER." In *Eurotransplant News-letter 53*, 10-18.

Czarniawska-Joerges, B. 1993. *The Three-Dimensional Organization: A Constructionist View*, Studentlitteratur:Lund.

Ekardt, H.-P. 1994. "Bautechnische Infrastrukturen. Beschreibung eines unvollständigen großen technischen Systems." In *Technik ohne Grenzen*, ed. I. Braun and B. Joerges (forthcoming).

Joerges, B. 1988. "Large technical systems: Concepts and Issues." In *The Development of Large Technical Systems*, ed. R. Mayntz and T. P. Hughes, 9-36. Boulder, Co.: Westview Press.

Joerges, B. 1992. "'Große technische Systeme': Zum Problem der Maßstäblichkeit und Vergrößerung in der sozialwissenschaftlichen Technikforschung." In *Jahrbuch Technik und Gesellschaft*, ed. Bechmann, G., Rammert, W., 34-60. Frankfurt/New York: Campus Verlag.

Joerges, B., Braun, I., 1994. "Große technische Systeme - erzählt, gedeutet, modelliert." In *Technik ohne Grenzen*, ed. I. Braun and B. Joerges (forthcoming).

Kubicek, H. 1994. "Steuerung in die Nichtsteuerbarkeit - Zur Entwicklung des großen technischen Infrastruktursystems der Telekommunikation." In *Technik ohne Grenzen*, ed. I. Braun and B. Joerges (forthcoming).

La Porte, T. R., ed. 1991. *Responding to Large Technical Systems: Control or Anticipation*. Dordrecht/Boston/London: Kluwer Academic Publishers.

Mayntz, R., Hughes, T. P., eds. 1988. *The Development of Large Technical Systems*. Boulder, Co.: Westview Press.

Mayntz, R., Schneider, V. 1988. "The Dynamics of System Development in a Comparative Perspective: Interactive Videotext in Germany, France and Britain." In *The Development of Large Technical Systems*, ed. R. Mayntz and T. P. Hughes, 263-298. Boulder, Co.: Westview Press.

Pichlmayr, R. 1986. "Von der Immunsuppression zur Immunmodulation am Beispiel der Organtransplantationen". In *Beobachtung, Experiment und Theorie in Naturwissenschaft und Medizin*, ed. Lüst, R. et al., 155-168. München.

Pruitt, T., Barett, W. 1991. "Corporate Virtual Workspace." In *Cyberspace. First Steps*. ed. M. Benedikt, 383-410. Cambridge, Mass.: MIT Press.

Rochlin, G. 1989. "Informal organizational networking as a crisis-avoidance strategy: US naval flight operations as a case study." In *Industrial Crisis Quarterly 3*, 159-176.

Rochlin, G., La Porte, T., Roberts, K. 1987. "The Self-Designing High-Reliability Organization: Aircraft Carrier Flight Operations at Sea." In *Naval War College Review*, (Autumn): 76-90.

Schoeppe, W. 1989. "Organization der Organentnahme und der Organtransplantation in Europa." In *Gesellschaft Gesundheit und Forschung*, Deutsche Stiftung Organtransplantation, 19-28. Neu-Isenburg.

Sherman, B., P. Judkin. 1992. *Glimpses of Heaven Visions of Hell. Virtuality and its Implications.* London: Hodder & Stoughton.

Sotto, R., 1993, "The Virtual Organization", Stockholm University, Dept. of Business Administration, Studies in Action and Enterprise, PP1993, no. 2, Stockholm.

3

Armies, Railroads, and Information: The Birth of Industrial Mass War

Arden   Bucholz

Armies can be viewed as examples of large technical systems (LTS) even though they are seldom studied from this perspective in the existing literature. Armies are indeed systems of machines and structures performing predictably complex, standardized operations, integrated with other social processes and legitimated by formal knowledge-intensive systems of rationality.[1] I argue that armies in World Wars I and II, Korea, Vietnam, the Falklands, Afghanistan and the Gulf War were second-order large technical systems. By second-order system is meant a recombination of two previously existing "classical" large technical systems (such as air traffic, railways or telecommunications systems) to form a third system (see Braun and Joerges, this volume).

In the specific armies described in this chapter, the systems that were recombined with the armies to produce industrial mass warfare were railways and information technology, specifically telegraph and telephone. The result was a high risk LTS that combined uncertainty of consequences with dangers to health, environment, identity and finances.[2] Its social and technological base was so huge it dwarfed even the largest civilian systems. After all, as soldiers and politicians are fond of reminding us, armies guarantee the existence of everything else.[3]

The difference between armies and other large systems is both technical and human. Armies are institutionalized not only on the basis of formal rationalities and modern procedures but also on more inclusive cultural principles. They combine the meticulousness and specificity of bureaucratic planning — goal oriented and rationalized — with an appeal to the senses and to the emotional and psychological force of the ancient and traditional drive for power. Because their ultimate goal

entails some degree of death, modern armies are among those industrial institutions that not only possess the "attributes of organization,
economic power and technical perfection but also generate 'spiritual
impulses,' social models and cultural ideals."[4]

In other words, armies are not only larger, they are different in kind
from civilian structures. Armies are large public monopolies, created
and maintained primarily to compete against each other on future battlefields. At the uttermost bounds of this competition, one or both of
the organizations will suffer some degree of death. As instruments of
death, a point which cannot be over-emphasized, and because of their
size and the extent and manner in which they command the loyalties
of their society, armies are unique among the large technical organizations of the modern world.[5]

Yet, outside the test environment of war games, staff exercises and
maneuvers, armies do not operate according to the designs and projections of their dominant managers.[6] On the contrary, in wartime they
often surpass the capacity for reflexive action by those responsible for
their design, management and operation. In the past, as Charles Perrow has argued, designers "could learn from the collapse of a medieval
cathedral under construction, or the explosions of boilers on a steam-
boat, or the collision of railroad trains on a single track." Twentieth
century societies have difficulties learning from nuclear plant accidents or chemical explosions.[7] The learning curve for war is even
flatter.

This paper describes the recombination of armies, railroads and information technologies in the quarter of a century before 1914 to create
industrial mass warfare. It deals with the internal transformation of
the five European great powers — Germany, England, France, Austria-
Hungary and Russia — as they each adopted deepfuture oriented war
planning processes with its consequences of high specialization of
knowledge and powerful external linkages to the European war system.
In other words, the application of modern transportation and communication processes transformed the size, space and time relationships of
traditional battle forces and helped to create the conditions of industrial mass warfare. These conditions produced casualties of twenty
million in the World War I and seventy million in World War II. The
birth of industrial mass war is described in four phases: the Prussian
invention of modern war planning (1864-1890), the initial diffusion of
the Prussian model in the period of counteractivity (1890-1905), the final diffusion of the Prussian model in the period of war crises (1905-
1913) and the July 1914 crisis which lead directly into the Great War.

The Origins of Modern War

Twentieth century war has its origins in the late nineteenth century. It began 1864-1890 with the invention of modern war planning in Prussia, its diffusion within Germany and the beginnings of technology transfer to Austria and France. The years 1890 to 1905 saw the continuing diffusion of the Prussian paradigm to Austria, France, Russia and England and the counteractive meshing of these states' philosophies of war, defense budgets, alliances and war plans. During this phase, isolated regional war systems began to be transformed into more integrated and tightly-coupled national ones. The last phase (1905 -1914) witnessed the final diffusion of the Prussian system and its integration via railroads into the now European-wide counteractive war system, both spurred on by six Cuban Missile type crises in the nine years culminating in the outbreak of World War I. In this phase, exceptionally rapid growth, the beginnings of a communications revolution, and the dominant impact of political developments became paramount. By the spring of 1914, five tightly coupled national systems based upon invariant, linear sequences were joined in a loosely coupled but highly interactive and time-dependent transnational system.

All large technical systems sometimes go astray. We are reminded of the accidents at Three Mile Island, Chernobyl and Bhopal. For armies the possibilities are slightly different. Instead of going astray, the resulting outcomes are often catastrophic. In the summer of 1914, after a quarter century of development, the largest second-order LTS in world history, the European war system, followed this spiral. By that time, the components of the system consisted of over twelve million men and one million railroad cars, stretched across three million square miles and was tied together by five national war plans based upon a core technology known as the military travel plan. With expectations of a war lasting 40 to 180 days, this huge transnational system culminated its period of radical reconfiguration by pitching itself directly into the first industrial catastrophe of global proportions, World War I.

Phase One:
The Prussian Paradigm (1864-1890)

Twentieth century war planning was invented and developed in the Kingdom of Prussia before 1863 and validated in a series of three wars from 1864 to 1871. These wars were unique in their size, space and time dimensions. In 1864, Prussia mobilized 65,000 men for 18 months and traversed a land area of a few hundred miles. In 1866, 280,000 men were mobilized and fought in 700 square miles. The war itself lasted scarcely two months. In 1870, more than 800,000 men were mobilized to fight in an area of over 1,000 square miles with field battles lasting six weeks. Thus in the six years between 1864 and 1870, the size of mobilization increased by roughly 1,500 percent, the land area increased 1,000 percent, and the time taken to mobilize was reduced by 400 percent.[8]

This magnitude of change was only possible by the application of the major technological motor of the nineteenth-century industrial revolution, the railroad. The only part of the war plan which could deliver continuous, reliable, predictable performance was the Military Travel Plan (MTP), which organized the railroads for mobilization. Generally speaking, the Prussians outnumbered the Danes 4 to 3, the Austrians 2 to 1, the French 3 to 1. In great measure, this was possible only because rail transportation delivered a large quantity of soldiers and weapons to a particular point at a specific time.

Europe looked at Prussian war-making in astonishment. Although it was not obvious at the time, observers were seeing the first deepfuture-oriented war planning process in world history. Deepfuture oriented meant the whole system was organized to focus and plan for five or ten years ahead. Even in its rudimentary form of 1871 this process had certain discernible parts. It was an organization based upon a division of labor by knowledge. The great general staff was divided and subdivided into components so that, in time, its tasks became coterminous with established areas of scientific and technical knowledge. The largest department, namely the land survey, was entirely based on scientific and technical knowledge undergirded by trigonometry. The most important department, mobilization, was dependent on the railroad section which was based on physics and mechanical engineering.

In the next twenty years, two changes occurred. First, Prussian war planning developed and expanded into the Second German Reich. In the process, deepfuture-oriented planning changed. Cycle time expanded. Procedures became more rigid. Manpower became more specialized. The organization as a whole[9] grew larger and the level of interactive complexity increased. Information requirements rose.

Secondly, by a process of technology transfer[10], some aspects of the Prussian system were taken up abroad. Austria, defeated in 1866, and France, defeated in 1870, both tried to learn it. Although geographic relocation was relatively easy, cultural diffusion was not and it met obstacles. Prussian war planning was expensive. It depended upon general economic conditions such as the extent and nature of the civilian railroad system. Regional geography, technological styles and legislation differed. New methods met opposition from traditional soldiers who argued that technology did not win wars: victory came through honor, bravery, self-sacrifice and the bayonet charge.

Phase Two:
Counteractivity to the Prussian Paradigm (1890-1905)

By the 1890s, a new development affected war planning, namely "counteractivity."[11] In other words, the Great Powers of Europe were caught up in a deadly network in which each reacted to one another. Its impact may be seen in four areas: philosophy, defense spending, alliances and war plans.

Philosophically, a climate of social Darwinism settled across Europe. Life was seen as a competition in which fit individuals, states and races survived and prospered, while the unfit weakened and died. Conflict, struggle and war were considered normal, natural and beneficial to the progress of civilization. This hostile science-based environment undergirded an emerging and violent world of ethnic politics.[12] It was not just that the world was thought to be more competitive, it was actively hostile.

Defense spending became counteractive. Each European great power watched its neighbors and reacted accordingly. The German defense bill of 1888 for the first time specifically considered French spending levels. The French had been anxious and fearful for some time. As the turn of the century approached, a new type of arms race began. Germany and France had started a conventional arms race in the 1880s. A high technology naval race began in 1898 with a German naval bill being more than matched by Britain during the ensuing Dreadnought era. The quick-firing French 75-mm gun, in service by 1898, signalled an artillery competition which continued from then on. Thereafter, submarines, more machine guns and other accoutrements were added. In an era of rapid technological change, defense budgets burgeoned against each other.

Two competing alliance systems were constructed. The triple alliance of Germany, Austria and Italy confronted the triple entente of France, Russia and England. Although many preliminary treaties, loans, staff talks and other negotiations had to be gotten out of the way, interactive agreements made uneven but steady progress. These agreements created networks so that if countries A and B allied and then country C attacked A, B would also fight, perhaps bringing in its allies D and E.

Finally, counteractive war plans were adopted. This was first apparent in Germany and France. As the result of a presumptive anomaly[13] uncovered in the test environment of a war game during the Moroccan crisis of 1905 to 1906, the Prussian general staff settled on a war plan which concentrated nearly two million men on the west front. Most of them were concentrated in the northwest quadrant where the highest density railroad system in Europe was to be the springboard for a huge rolling offensive against France, scheduled to end the west front campaign in about forty days. The offensive was later described as the most ambitious project ever undertaken for controlling the immediate future of so many people. All European war mobilization plans — the French Plan 15, Russian Plan 19 and Austrian plans R and B — followed suit.

Following the Prussian paradigm and as a result of size, space and time factors in mobilization, these plans were substantially dependent on their railroad component, the Military Travel Plan.[14]

The laws of physics and mechanical engineering impose certain operating rules on railroads. One is uniform speed. It is more efficient to keep trains moving continuously at as high a speed as possible. Thus the fewer stops, slow downs and speed ups, the more efficient the railroad. Traffic is organized in "correct speed flights" following the principle that greater volume is achieved when all trains move at equal speed.

A second operating rule is interdependency. Almost everything in an active railroad network influences rather immediately and directly everything else. For example, one late train may cause following trains on the same track to slow down, proceeding trains to be shunted aside, and connecting trains at junctions to wait. All of this action radiates out in many directions from a single moving point.

A third operating rule is discipline. Rail transportation, in contrast to road transportation, is a highly disciplined movement. Uniformity and interdependency require 100 percent mechanical synchronization of speed and time. To maintain this discipline requires huge and steady amounts of information. Everything must be predictable and follow pre-arranged, written directives. Discipline depends on timely and accurate information.[15]

These principles — uniformity of speed, interdependency of parts and discipline of movement — result in a particular kind of organization, namely a system nested hierarchically. It is based on blocks, lines, divisions and regions. In the block, one train at a time is allowed. As one block receives a train, the block behind protects it and the block ahead is asked to receive it (see Usselman, this volume). Electric signals alert receiving and sending blocks 12 to 15 miles in all directions. Above the block is the line. In this unit, space and time are measured by line capacity, i.e. the number of trains which can pass through in a given time. To achieve the continuous movement which is the optimum goal of railroads, the maximum number of movements has to be planned for in advance. The goal is continuous process or flow-through technology. Continuous process requires control based on continuous information. The result of this flow-through technology can be seen in the final German war plan which called for 560 fifty-four car unit trains to cross the Hohenzollern Bridge at Cologne each twenty-four hours for eleven days.

Preliminary work by this author suggests that similar developments occurred outside Germany also. French War plans at the Chateau de Vincennes in Paris, Austrian plans in the Kriegsarchiv in Vienna, English documents in the Public Record Office in London, and Russian materials such as those of Zayonchkovski, Ushakov and Ronzhin reveal similar developments. In spite of the cultural diversity confronting similar technologies, lost wars spurred on technology transfer. England after the Boer War and Russia after its disastrous confrontation with Japan reformed their war planning processes.

The result was that the principles of physics and the discipline of mechanical engineering specified the basic operating rules of the Military Travel Plans. Gradually these axioms — uniformity, interdependency and discipline — assumed a dominant place within European war systems. They affected organizational structure and ethos. As railroads became the core technology of war planning, the systematic application of organized knowledge applied to the practical skills of war resulted in the creation of a new kind of organization and personnel. Both were science- and information-based as well as engineering oriented.

Phase Three:
Crisis and Final Diffusion (1905-1914)

By 1905, the third and final phase of the evolution of industrial mass war had begun. This phase comprised a series of war crises — a total of six crises during the next nine years — averaging one every eighteen months. There were the First Moroccan (1905), Bosnian (1908), Second Moroccan (1911) crises, followed by the Balkan Wars of 1912 and 1913 and finally the July crisis of 1914. Each crisis, which foreshadowed in various ways the Cuban Missile Crisis of 1962, tightened the war planning processes. Like a bundle of micro-computer chips, each war planning system became more tightly coupled. Although each crisis contained unique elements which did not all come together until July 1914, their impact was cumulative. Gradually the tension and stress levels of the transnational system increased. In addition, as the Great Powers became engaged in a telegraphic and telephonic communications revolution of the first magnitude, the European war system took on some attributes of a sensitive hair-trigger, increasingly in danger of being pulled by nervous rulers, spurred on by their technical advisors. It is estimated that in the final 72 hours before war declarations in the summer of 1914, more than 2,600 telegrams, many encoded, were exchanged between the Great Powers.

The first Moroccan affair from 1905 to 1906 was almost a non-crisis. Although the German general staff probably advised war against France, and British warships were recalled from the Far East, the European powers were unprepared for war. Russia was bogged down in Asia (fighting the Japanese) and at home by a revolution. Austria was uninvolved. A surprised France, fearing defeat and another Paris Commune, backed down completely. This crisis, however, caught the attention of Europe like a red flag. England hurried to upgrade and modernize its military planning system. Staff talks increased between England and France. Russia, buoyed by French loans accompanied with specific instructions, put on the drawing boards a massive program of railroad building along its northwest frontier.

The Bosnian crisis from 1908 to 1909 also had its non-players. France, especially, was uninvolved and refused to support its ally Russia. England was on the sidelines. However, within the triple alliance, the crisis dramatically altered German-Austrian military ties by changing the defensive alliance of 1879 into an offensive treaty. Like France in 1906, Russia was unprepared for war and, confronted by two German ultimatums, backed down — with consequent loss of face internationally — and stepped up preparations internally. Historians have argued that this was a dress rehearsal for 1914.

The second Moroccan crisis of 1911 had its non-participants as well, for example Russia. However, England emerged as a major player, indicating to France its support in case of war. France, like Russia in 1909, backed down militarily but began final transformation of its war plans and processes.

The Balkan Wars, 1912 and 1913, resulted in several more ultimatums and the first trial mobilizations as both Austria and Russia assembled armies of several 100,000 men each for many months in the field. In late autumn 1912, general staff chiefs in both St. Petersburg and Berlin voiced sentiments that war was coming and the sooner it came the better. By 1913 European technical war plans were loaded and the organizational control system was cocked.[16] These plans and control systems were largely invisible and little known to political actors.

The Impacts of Railroads

By this time, the general implications of deepfuture oriented war planning were substantial. All European war plans were achieved by the use of railroads, the dominant mechanical form of large scale transportation, with the help of the telegraph and telephone as electrical forms instrumental to railroads. It is probably not an overstatement to say that the technology of railroads dominated European war planning. For European general staffs, this had six general implications. [17]

One impact was increased performance and an intensification. The year 1905 heralded the beginning of an era in which military expectations began to assume that all cogwheels would operate without "play." The disappearance of margins, or a narrowing of tolerances, has been used as a paradigm for the machine in general. Up to a point, railroads both forced and allowed a more precise interaction between the separate parts of armies. Napoleonic armies always contained a great deal of play. Even Moltke's corps could arrive late. By Schlieffen's day, a great deal of play had been squeezed out and by 1914 it was further reduced, both in the war plans and, equally important, in the minds of the planners. Close tolerances of time, space and size were assumed.

A second impact of the railroad in war planning was that mechanical regularity began to triumph over natural irregularity. The replacement of animal and human power by steam power in one stage of the war plan assumed the guise of a guarantee for the whole plan. The natural irregularities of the terrain were replaced in the planner's mind by the sharp linearity of the railroad. A machine ensemble had injected itself between the railroad and the land: war planners now began to think about the land as it was filtered through the machinery.

A third impact was a reorganization of traffic according to the hardware technology of the railroad. If the German plan of 1905 was based upon considerations of European rail density and the Moltke plan of 1914 hinged around the capture of a single rail traffic point, this was the rule rather than the exception. For France (Plan 17), for Austria-Hungary (Plans B-1, B-2 and R), and for Russia (Plan 19), all plans depended upon the placement and usage of railroads. European general staffs charted the progress of their war plan using railroad maps which described time and size instead of distance. These industrial flow charts measured the execution of the war plan on the basis of work capacity and timing, just like a factory production line.

This is clearly visible in Figure 1, the German War Transportation Plan Map for Section 4, Danzig Sector, April 1, 1913 — April 1, 1914.

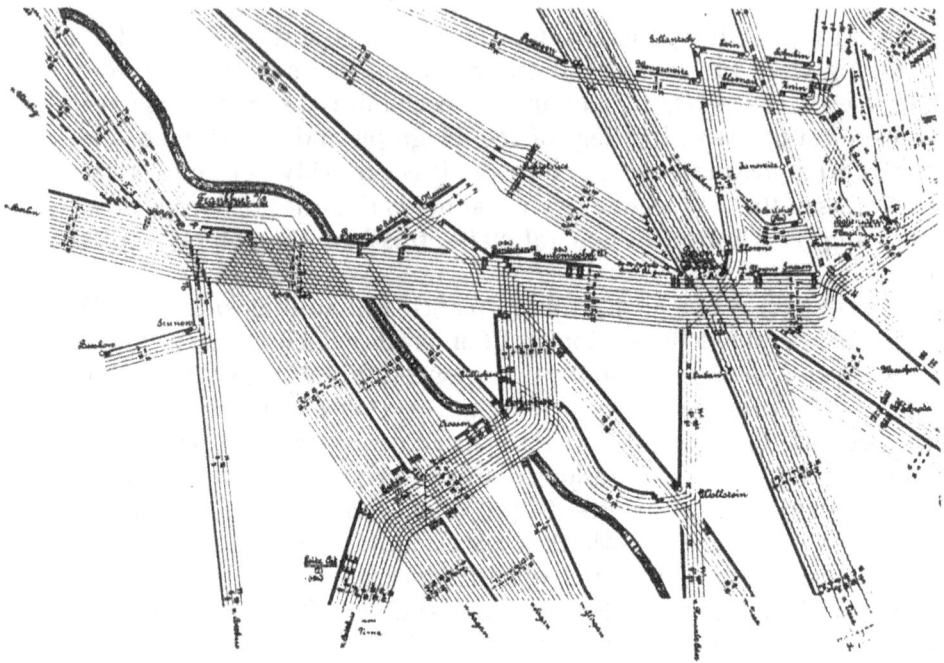

FIGURE 1 This map, produced in the Railroad Section (the core technology) of the German General Staff, is the War Transportation Plan Map for Section IV, Danzig Sector. It illustrates the flow of traffic east and south of Frankfurt am Oder toward the Polish and Russian borders, in execution of the German war plan. The map measures space and size in terms of time. It illustrates work capacity, figured mathematically and described visually by the thickness of the lines. (*Source*: U.S. National Archives Microfilm, Groener Papers, roll 19)

The fourth impact of the use of railroads in war plans was that the newly created European general staffs' ethos was shaped by their confrontation with technology. Bureaucracies within European armies became some of the most modern organizations within traditional societies. The metamorphosis from patrimonial to rational management resulted in a shift of power from rulers to subordinates and from generalists and dilettantes to technical experts.[18] As European armies became railroad-dependent, the task of size, space and time coordination created a new kind of officer. The general staff officer became the model of consistent, dependable, technical performance: he was interchangeable. His goals were functional reliability and high work capacity. The general staff ethos derived from its technical core, namely railroads.[19]

Fifth, after prolonged growth, large organizations such as armies often acquire momentum. By this is meant that they have many technical and organizational components, possess direction and display a rate of growth suggesting velocity.[20] Load factors, in this case the ratios of average peacetime output to war-mobilized maximum output, and load curves tracing these changes over time began to suggest trajectory. This is particularly true of military organizations, those large public monopolies created and maintained primarily to compete against each other on future battlefields.

In mobilization, the high momentum system of Germany confronted the medium velocity one of France and the low momentum one of Russia. In other words, the German system could mobilize and concentrate a huge force in about fifteen days. France took longer and Russia took much longer. Both vested interests and fixed assets determine the trajectory and velocity of momentum. Although European armies had some appearances of the closed system typical of autonomous technology — inner dynamic, managerial decisions to reduce uncertainty and increase capacity — these appearances were deceptive. In reality, prevailing political pressures and values, as well as governmental regulations that expressed these, matched the internal dynamic of the military systems. By 1914, technology and politics moved in the same direction, increasing the velocity finally achieved.

Sixth, there may have been a relationship between the planning processes used within the railroad sections of the European general staffs and the transportation method called "linear programming." This method undergirds critical path theory, a management technique widely discussed in the 1960s. It assumes that in any large project, a small number of operations control the project completion time. These tasks form an identifiable chain or "critical path" through the larger project. In German war planning this method may have been taken a step further, with the use of the cycle paradigm in contrast to the ladder paradigm in the evolution of the military travel plan and even for the larger war plan itself. In such a process, technology tends to produce a homogenization or narrowing of choices and possibilities. This technique, which allowed German planning to evolve in incremental stages, as one generation of plans succeeded another, may have given German war planning from 1914 to 1941 the same competitive advantages among European armies that Japanese manufacturing cycles are said to have achieved in the past decade over American methods.[21]

Political Imbalance

Meanwhile, outside the general staffs there was a serious problem: the political component of the European war network had fallen behind the military component, resulting in an imbalance. The historian of technology Hughes defines this type of problem as a reverse salient.[22] Anthropologists sometimes call it cultural lag.

One aspect of this imbalance was that European states lacked modern decision-making bodies. Assessment tended to be concentrated among a handful of persons, with few routines to discuss and coordinate between these persons and technical experts. As the crisis developed, therefore, there was no management, no consideration of options, no weighing of alternatives. The means of making war, weapons, transportation and communications had made the jump into the twentieth century, while the European political system operated in a nineteenth-century cabinet atmosphere where the whim of the monarch or prime minister called the tune.

A second aspect was a knowledge gap between the military technicians and the political actors. A significant degree of incomprehensibility accompanied the outbreak of World War I. Information about the war plan was restricted to those who had a need to know. The same was true of intelligence. European diplomats and monarchs lacked clear understanding of the various kinds of mobilizations. German mobilization had eight stages but this information was unknown even to the German chancellor. Germany did not communicate this to Austria. Austrian mobilization needed a minimum of sixteen days, which was not known to German political actors.[23] A transnational technical system existed beneath the level of political actions; however, this sytem was in varying degrees unknown and unsuspected by political actors.

A third aspect of imbalance was that the counteractive nature of the war plans exerted great pressure upon political leadership. Separate state war plans were tightly-coupled systems with time dependent processes and invariant sequences, while coupling between alliance partners was looser, but political leaders were completely unfamiliar with these tensions. Each war plan was measured in stages, hours and days and all were technical devices dependent on the railroads' mechanical engineering. Complex notification procedures meant communication processes were lengthy. Because of the public, visible and complicated maneuvers needed to get mobilized armies into the field, each state was anxious about other's actions. Preliminary steps by one were considered ominous or dangerous by others. Information about such preliminary moves was poor at best but "guaranteed" under the circumstances to be examined, misunderstood and reacted to by opposing powers.

A final aspect of the crisis was the effect of the overall milieu, including the media and urban crowds, on political leadership in Berlin, Paris, St. Petersburg, Vienna and London. Monarchs or chiefs of state, however popular they may have wanted to be, were unaccustomed to large crowd stress. During the final days of July 1914, governmental deliberations changed from private and secret to public and publicized. Any publicity was unnerving, but hourly newspaper editions and massive, ominous crowds, especially considering the underlying social Darwinist milieu and nearly decade-long crisis environment, probably moved leaders in a war-like direction.

By 1914, political leaders as well as intellectuals had been influenced by Nietzsche and the "warrior mentality." As redefined by Michel Foucault, there are two kinds of truth: one obtained by inquiry, survey and inquest, the other reached by ordeal, test or trial. It was this latter kind that tempted a handful of the European leadership elite in 1914. In this case, truth does not belong to the order of that which is but rather to that which happens. It is not recorded but aroused. In other words, action becomes the occasion for the production of a particular kind of truth. With this characterization, Nietzsche had added the finishing touches to social Darwinism, urging forward to action and climax. After forty-three years of peace in Europe, some leaders were ready to take the plunge, to reach the known boundaries and beyond in hopes that the experience might yield a new kind of truth and the birth of a fresh, chastened and more vigorous civilization.[24]

The July 1914 Crisis

The July 1914 crisis began as an act of traditional nineteenth century terrorism, namely the grotesque and almost comic opera assassination of visiting royalty. It ended as the modern mechanical launching of the first industrial catastrophe of global proportions. The crossing point from traditional to modern came with the 48-hour Austrian ultimatum to Serbia of 6 p.m., Thursday, July 23, 1914. With the ultimatum, the time clock of the war plans began to tick.

In this science-based social Darwinist finale, every action was measured in terms of time and space that were quantifiable. Officers thought of them in terms of numbers. As General Joffe said on July 31, every delay of 24 hours in calling up reserves meant an initial abandonment of 15 miles of territory.[25] As opponents began to mobilize against each other, all European general staffs felt the urgency of time. There were points, different for each state, at which counteractive mobilizations became triggered technologically, that is, by knowledge or suggestion of opponents' bureaucratic and operational moves. When news of Russian partial mobilization reached Berlin on July 26, German war planners probably ordered into effect the first two stages of their war plan, "state of security" and "political tension." Because of the interlinked design of each war plan, various points along the trajectory — once reached — triggered additional engagements. When orders were given for the partial mobilization of four Russian military districts on July 25, it was not clear to the decision-makers that this would impede subsequent general mobilization. Other points, once passed, could not be reversed but had to proceed in forward position. The Austrian Plan B, once past sixty hours, was irreversible: to redeploy those troops against Russia, they had first to be transported to the Serbian end of the line and then returned.

The impact of military technology on political decision-making was in reality the impact of the specialist on the dilettante. In other words, it was the impact of more knowledge over less or no knowledge. In worse-case war scenarios, political leaders deferred to military planners whose plans appeared absolute and certain because they were based on the mathematics of engineering. Although this was relatively unclear in peace time, by the end of the July crisis it had become highly evident. The result was the clearest in Germany.

On July 28, 120 hours after the Austrian ultimatum to Serbia, German General Staff Chief Moltke sent a secret situation assessment to his chancellor. Moltke stated that Germany's options were now gone. As a result of Austrian and Russian mobilizations, the direction and movement of events were out of control. Only the speed could still be determined. The next 72 hours were crucial. Germany could not allow these to slip away without acting or the most grievous consequences would befall her.

Finally, on the afternoon of August 1 after the mobilization documents had been signed and the implementing orders were being processed, the German Kaiser Wilhem II changed his mind. Having heard from his ambassador in London that there was apparently some prospect of English neutrality, Wilhelm II ordered a halt to the west front deployment and a redeployment of all forces against Russia. Moltke finally convinced the Kaiser that this was impossible. The final war plan, the result of a year's work, could not be altered. The result would be to send a swarm of disorganized troops to the east front while at Germany's rear there was a fully war-mobilized French army. Moltke said he would take no responsibility for this chaos.

Moltke and other military leaders telegraphed London, saying that the deployment could not be stopped for technical reasons.[26]

Acknowledgements

Portions of this chapter were presented at the conference "Technological Development and Science in the 19th and 20th Centuries," University of Technology, Eindhoven, the Netherlands, November 1990, and at the annual meeting of the Society for Military History, sponsored by the U.S. Marine Corps Command and Staff College, Fredericksburg, Virginia, April 1992.

I am indebted to Martyn Bakker, Peter Kores, Gerd Verbong, Dennis Schowalter and Allan Capps for comments. I would also like to express thanks to my colleague Dr. Walter Boston and his International Studies Program for funding some of the research travel; thanks also to my friends and colleagues Drs. Neil Johnson, Kempes Schnell, John Kutolowski and Steve Ireland for reading and commenting on various drafts.

Notes

1. This formulation builds on Joerges' definition of a technical system, see Joerges 1988, pp. 23-24.

2. See Joerges 1988, p. 10.

3. As Nicolo Machiavelli wrote, "The chief foundations of all states, new as well as old or composite, are good laws and good arms; and as there cannot be good laws where the state is not well armed, it follows that where they are well armed they have good laws." Bartlett, 1938, p. 1023.

4. Buddensieg, 1984, p. 2.

5. Bucholz, 1991, p. 3.

6. See Joerges, 1988, p. 26, who discusses large technical systems in general. His comments would seem to apply to armies as well.

7. Perrow, 1984, p. 12.

8. Craig, 1964, passim; Howard, 1969, passim; Bucholz, 1991, ch. 1.

9. Bucholz , 1991, ch. 2. Armies as organizations within the state system of the past century can be approached in a variety of ways. A seminal article is Pye, 1967, pp. 69-89; a second fundamental approach is provided in Galbraith, 1974. For a provocative discussion of contemporary, complex and knowledge-based organizations, see Perrow, 1984.

10. On technology transfer, see Kenwood and Lougheed, 1982; Hugill and Dickson, 1988; Headrick, 1988; Westney, 1987; Ralston, 1990. A recent article argues that (1) technology diffusion follows an S-shaped pattern: innovation is

taken up slowly at first, then by a rapidly increasing number and finally by a few laggards; (2) application is easier if management and the work force are well-educated and operate in an institution in which they change jobs every three to five years; (3) technologies are most easily applied by institutions which have developed a "culture of innovation," of which education is the foundation; and (4) research into innovation in England, 1945-1983, found that over 66 percent originated in the fields of engineering and chemicals. *The Economist*, 11-17 jan. 1992. Some of these ideas could be usefully applied to European armies in the quarter century before 1914. For insights into the opposition of traditional soldiers to new technology, see the articles of Michael Howard and Stephen Van Evera in Miller, 1985.

11. Ritner in "The Society of Space" calls situations of counteractivity "weave problems" and argues that they are not susceptible to cause and effect logic but must be dealt with by mutual dependence analysis. Toffler, p. 174.

12. Bucholz, 1991, ch. 2.

13. Presumptive anomalies occur when assumptions derived from science indicate that under some future conditions the conventional system will fail or function badly or that a radically different system will do a much better job. In this case the presumptive anomaly was that indoor war gaming revealed the possibility that the assumed conservative attack might, under certain conditions, fail and that a radically different strategy might succeed. Hughes, 1987, p 25; Kern, 1983, p. 285; Bucholz, 1991, pp. 195-210.

14. Bucholz, 1991, chs. 5 and 6.

15. Ibid.; See Marvin, 1988 and Yates, 1989.

16. Bucholz, 1991, ch. 6.

17. See Schievelbusch, 1985, pp. 16-26. Provocative insight on the impact of technology in creating new modes of thinking about time and space is contained in Kern, 1983.

18. Bucholz, 1991, chs. 6 and 7.

19. Ibid. One can wonder to what extent core technology planners within European general staffs (that is, those associated with the Military Travel Plan and railroads) depended upon mathematical relationships illustrated through visual images. If so, they were thinking in terms of performance curves, for example efficiency versus load or stress versus strain. In other words, they envisioned their railroad systems' performance characteristics under various specified wartime or worse case scenarios. See Ferguson, 1992, p. 148.

20. See Hughes, 1987, p 76.

21. On critical path theory , see Horowitz, 1980; Baker, 1974; Bowman, 1966, pp.475-478; Wiest and Levy, 1977; Gomory, 1990, p. 191.

22. Reverse salients are technical or organizational anomalies resulting from uneven evolution of a system: progress on one front is accompanied by stagnation on others. Hughes, 1987; Joerges, 1988, p. 13.

23. See Herwig, 1984, p. 89-94. This suggests a breakdown in communication between the two groups similar to that between the "expert sphere" and the "user sphere" described by Pacey, 1983, pp. 48-51.

24. See Hale, 1940; Miller, pp. 269-273.

25. Kern, 1983, p. 273.

26. Bucholz, 1991, p. 311-2. Had C.P. Snow's two worlds of scientist and humanist split apart? No. Rather, as Hans Delbrueck wrote in late October 1918, Germany had only *Fachmenschen*, technical specialists, and even they did not appreciate their power. Bucholz, 1985, p. 108. Drucker argues that management fits neither of Snow's categories, rather it is a technology. 1989, p. 231.

References

Baker, Ken. 1974. *Introduction to Sequencing and Scheduling*. New York: John Wiley.

Bartlett, 1938. *Familiar Quotations*. 11th ed. Boston: Little Brown.

Bowman, Edward H. 1966. "Production Scheduling by the Transportation Method of Linear Programming." In *Readings in Production and Operations Management*, ed. E.S. Buffa, 475-478. New York: John Wiley.

Bucholz, Arden. 1985. *Hans Delbrueck and the German Military Establishment*. Iowa City: University of Iowa Press.

_____ 1991. *Moltke, Schlieffen and Prussian War Planning*. Oxford and New York: Berg Publishers, St. Martin's Press.

Buddenseig, Tillman. 1984. *Industriekultur: Peter Behrens and the AEG*. Cambridge: MIT Press.

Child, John. 1984. *Organization: A Guide to Problems and Practices*. 2d ed. London: Harper and Row.

Craig, Gordon. 1964. *The Battle of Koeniggraetz*. Philadelphia: Lippincott.

Drucker, Peter. 1989. *The New Realities*. New York: Harper and Row.

"Innovation: The Machinery of Growth." *The Economist* 322, No. 7741. (11-17 Jan. 1992): 17-19.

Ferguson, Eugene S. 1992. *Engineering and the Mind's Eye*. Cambridge: MIT Press.

Galbraith, John Kenneth. 1974. *The New Industrial State*. 2d ed. London: Penguin Books.

Gomory, Ralph. 1990. "Of ladders, cycles and economic growth." Scientific American 262, No.6 (June): 140, 180-182.

Hale, Oron. 1940. *Publicity and Diplomacy: with Special Reference to England and Germany, 1890--1914*. Cambridge: Harvard University Press.

Headrick, Daniel. 1988. *The Tentacles of Progress: Technology Transfer in the Age of Imperialism, 1850-1940*. New York: Oxford University Press.

Herwig, Holger. 1984. "Imperial Germany." In *Knowing One's Enemies: Intelligence Assessment Before the Two World Wars*, ed. E. R. May, 89-94. Princeton: Princeton University Press.

Horowitz, Joseph. 1980. *Critical Path Scheduling*. New York: John Wiley.

Howard, Michael. 1969. *The Franco-Prussian War*. New York: Collier.

Hughes, Thomas P. 1987. "The Evolution of Large Technical Systems." In *The Social Construction of Technological Systems. New Directions in the Soci-*

ology and History of Technology, eds. Bijker, W.E., Hughes, T.P. and Pinch, R. J., 51-82. Cambridge: MIT Press.

Hugill, P. J. and Dickson, D. B., eds. 1988. *The Transfer and Transformation of Ideas and Material Culture*. College Station: Texas A & M University Press.

Joerges, Bernward, 1988. "Large technical systems: Concepts and issues." In *The Development of Large Technical Systems*, eds. R. Mayntz and T. Hughes, 23-24. Boulder, Colo: Westview Press.

Kenwood, A.G. and Lougheed, A.L. 1982. *Technical Diffusion and Industrialization Before 1914*. New York: St. Martin's Press.

Kern, Stephen. 1983. *The Culture of Time and Space*. Cambridge: Harvard University Press.

Marvin, Carolyn. 1988. *When Old Technologies Were New: Thinking About Electric Communications in the Late Nineteenth Century*. New York: Oxford University Press.

Miller, Steven. 1985. *Military Strategy and the Origins of the First World War*. Princeton: Princeton University Press.

Pacey, Arnold. 1983. *The Culture of Technology*. Cambridge: MIT Press.

Perrow, Charles. 1983. *Normal Accidents: Living with High-Risk Technologies*. New York: Basic Books.

Pye, Lucian. 1967. "Armies in the Process of Political Modernization." In *The Role of the Military in Underdeveloped Countries*, ed. J. J. Johnson, 69-89. Princeton: Princeton University Press.

Ralston, David. 1990. *Importing the European Army: The Introduction of European Military Techniques and Institutions into the Extra-European World*. Chicago: University of Chicago Press.

Schievelbusch, Wolfgang. 1985. *The Railway Journey: The Industrialization of Time and Space in the 19th Century*. Berkeley: University of California Press.

Scott, W. Richard. 1981. *Organizations: Rational, Natural and Open Systems*. Englewood Cliffs: Prentice Hall.

Toffler, Alvin. 1980. *The Third Wave*. New York: Bantam Books.

Weist, Jerome and Levy, Ferdinand. 1977. *A Management Guide to PERT/CRP*. Englewood Cliffs: Prentice Hall.

Westney, Eleanor. 1987. *Imitation and Innovation. The Transfer of Western Organizational Patterns to Meiji Japan*. Cambridge: Harvard University Press.

Yates, JoAnne. 1989. *Control Through Communication*. London: John Hopkins University Press.

4

Multinationals in Transition: Global Technical Integration and the Role of Corporate Telecommunication Networks

Volker Schneider

> The modern corporation is a far cry from the small workshop or even from the Marshallian firm. The Marshallian capitalist ruled his factory from an office on the second floor. At the turn of the century, the president of a large national corporation was lodged in a higher building, perhaps on the seventh floor, with greater perspective and power. In today's giant corporation, managers rule from the top of skyscrapers; on a clear day, they can almost see the world. (Stephen Hymer, 1972)

In recent years we have been confronted with an amazing paradox: an increasing number of multinational corporations (MNCs) such as GM, IBM or Siemens are decentralizing their organizational structures and trying to gain more flexibility in their production systems. Some observers conceive these changes as symptoms of a more fundamental transformation in production and industry structures. Some scholars herald the end of mass production (Piore and Sabel 1984), while others promise the end of hierarchy (Peters 1992). There seems to be a general optimism in the literature that expects a new age of markets marked by more downsized economies and more flexible socio-economical settings.

At the same time, however, business firms have continued to become larger in size and more global in scope. The turnover of biggest companies is reaching dizzy heights, surpassing the GNP of an increasing number of nations, and the operational scope of multinational corporations almost covers the surface of the earth. Almost all MNCs try to market their brand-name products all over the planet.

The main argument of this chapter is that greater organizational decentralization and higher production flexibility are not the end of hierarchy and mass production but just a reconfiguration of their form. Flexible specialization does not decrease production scales and organizational decentralization does not eliminate central control. The chapter's core hypothesis is that central control via the traditional type of hierarchies is being substituted by control via technical systems, integrating MNCs, their subsidiaries and their supplier firms. Increasing global technical integration of administrative control and production processes solves the paradox of organizational decentralization by concentrating on fewer supplier firms (single sourcing) and outsourcing strategies. By outsourcing is meant that corporations reduce their activities to "core competencies" and shift peripheral tasks to subcontractors — often at different places in the world.[1] MNCs shrink their legal cores by shifting more and more elements of the value chain to outside suppliers. At the same time, however, these supplier firms become tightly integrated and synchronized with the core production systems of their parent company. Suppliers are reduced — despite legal autonomy — to mere technical satellites in a global production system.

The chapter shows that this form of integration has been made possible by communications and information technologies developed over the past thirty years. I will show that the shaping of industrial organization through communication technologies is a relatively new phenomenon (though it has happened several times in history). The final section on recent history reveals how the current organizational reconfiguration of MNCs goes hand-in-hand with higher forms of technical integration on the basis of specialized corporate telecommunications networks, thus transforming MNCs and their satellites into global industrial production systems.

Global Corporations and the Development of Large Technical Infrastructures

The organizational growth of MNCs, whose global technical integration in production is largely based on the growth of global transport and telecommunications networks, offers an exemplary insight into the vital importance of large technical infrastructures in the development of advanced industrial societies. Just as the industrial revolution and societal modernization are unthinkable without the growth of modern transportation and telecommunications systems (Mayntz 1988: 234), the current global firms would be inconceivable without the turbo jet, transoceanic cables, communication satellites and international computer networks (Chandler 1986, Mulgan 1991).

The study of global corporations provides an interesting perspective on economic and political globalization or, more generally, on the specific interworking of societal development dynamics, such as social differentiation, institutional integration and the growth of communications infrastructures. MNCs provide a societal locus where all these dynamics work together:

- MNCs are complex social configurations displaying functional differentiation at all levels. They internalize micro differentiation at the task level à la Babbage, meso differentiation à la Durkheim, and division of labor among countries à la Ricardo.
- MNCs are continents of hierarchies in oceans of world markets. They display complex patterns of institutional integration and modes of governance. However, many MNCs are decentralized to such a degree that some intraorganizational relations in the global hierarchies are changed into market-like structures.
- Modern MNCs are nested Large Technical Systems (LTS) which provide technically integrated production and distribution systems interfacing with global technical infrastructures such as transportation and telecommunications.

It is important to note that the three complexes above are not mere analytical perspectives but dimensions of a unique social reality. Functional differentiation is closely related to patterns of institutional integration and both are based on technical infrastructures. Communication technologies are particularly important prerequisites for facilitating differentiation and institutional integration.[2]

By highlighting the evolution of MNCs, we can study the interworking between differentiation, institutional integration and communication in detail. We see that the division of labor and the growth of markets and hierarchies were inseparably linked to the parallel growth of global communications infrastructures, and that new decentralized organizational forms of MNCs could only be designed on the basis of advanced global communications structures. The main theses are thus that:

- organizational differentiation, the growth of corporate hierarchies and the spread of technical communications and transportation systems are highly interdependent,

- the current form of decentralized globalization only became possible through the parallel emergence of a global communications and transportation system,
- organizational decentralization is paralleled by tighter technical integration of production systems.

Corporate Growth and Communications Revolutions

Multinational firms have a long history. Long before the modern MNC emerged, institutional arrangements were invented to put private economic cross-border activities on more stable grounds (Ells 1972). Examples of early merchant associations can be found in ancient Greek and Roman times, but the medieval merchant societies and the trade houses characteristic of early capitalism are better known.[3] In the first half of the sixteenth century, the south German Fugger dynasty, for example, owned and managed trade agencies in a dozen kingdoms and operated mines scattered over a wide geographical area. The control structure of this enterprise was based on intra-family kinship relations, although some early forms of "joint ventures" between families were also used (e.g. with the Hungarian Thurzo family). More recent historical examples of non-personal organizations are the British colonial trading companies (Smith 1789: 620-640).

Many of the late medieval and proto-capitalist trade companies contributed to an early accumulation and concentration of economic resources within Europe. However, none of these companies were the forerunners of the modern multinational corporation (Hymer 1972). Rather the ancestors of the modern MNC were the small workshops in the late eighteenth and early nineteenth century, whose success was based on a specific combination of production technologies, factory hierarchy and market exchange. The underlying organizing principle of factory hierarchy was, on the one hand, institutional differentiation between owner and worker (private ownership of production technology, employment contracts implying control by authority) and on the other hand, technical division of labor among workers. The factory hierarchy was a new method of coordinating differentiated activities. In contrast to medieval guilds and corporations — where production and exchange were hierarchically structured through detailed prescriptions of applied skills, quality and prices — industrial capitalism consisted of a market hierarchy combination in which only the technical division of labor within the factory hierarchy was consciously planned. The transactions between factory hierarchies were regulated by decentralized and unconscious market processes.

The evolutionary superiority of the new technical and institutional arrangement was based on a specific mechanism of productivity growth. An increasingly refined division of labor that was hierarchically planned and combined with continuously improved production technologies led to an unprecedented increase in production output. Exchange processes in competitive markets put pressure on firms to reinvest in production technology and human skills, which improved the productivity even further. Continuous accumulation, progressive division of labor and technical progress produced a self-stimulating growth spiral.

As the production volumes increased, so did the size of firms. However, modern industrial enterprises became larger not merely by expanding their production systems but also by internalizing and interfacing with raw material extraction, distribution networks (sales and purchasing offices), transportation systems and later, research and development laboratories (Chandler 1980:397). The coordination of these increasingly complex multi-unit enterprises required specialists and specialized technologies. A new profession called management emerged.

Both telecommunications systems and major innovations in transportation were significant examples of technological integration in the nineteenth century that made internalization possible. Telegraphy, railroads and the steamship enabled the integration of geographically dispersed production and distribution units. This is an interesting corollary to the emergence of the multinational trading associations in the fifteenth and sixteenth century, which were based on a communications revolution, too. It was made possible through the diffusion of industrial paper production,[4] the invention of double-entry bookkeeping[5] and the emergence of the first postal systems in Europe.[6]

Growth of Firms in the 1800s

The growth of industrial firms in the nineteenth century was largely conditioned by the growth of large technical infrastructures. The industrial revolution resulted not only from innovations in power and energy use but also from revolutionary developments in communications and transportation.[7] Prior to these changes it "probably took longer to get a message from York to London, for instance, in 1750 than it had taken in Roman days, for the possible improvement in the breeding of horses was more than offset by the very definite deterioration of the roads" (Albion 1932: 719). Without the communications revolution of the last century, mass production and distribution would not have been possible (Chandler 1984).

Since new forms of economic differentiation and integration are thus closely related to the emergence of specific socio-technical infrastructures, it is no coincidence that the first large-scale industrial firms and the first industrial MNCs appeared only in the second half of the last century (Chandler 1977, 1980, 1984, 1987, 1989). Corporations expanded by growth, merger and vertical integration. They became multinational either by forward integration, i.e. by establishing distribution channels abroad, or by backward integration, i.e. by gaining control of the provision of raw materials. Firms' foreign activities were mainly limited to distribution channels up to the late nineteenth century, but from the 1880s on, a number of firms also situated manufacturing plants abroad. Most of these early MNCs were European (especially British and German) and U.S. companies.

Organizational integration across nations and continents required a number of logistical prerequisites: economies of scale through mass production supposed high capacity utilization. This called for a constant flow of materials which could not be guaranteed through traditional means of transportation (based on animals, wind, etc.). Thus, the parallel growth of firms and large technical infrastructures contributed not only to an increase in the internal efficiency of the firm but also externally to an expansion of the scope of its markets. The process became self-stimulating — expanding integrated firms in expanding markets accumulated increasing resources. This led to rising fixed costs and falling average costs, reinforcing the pressure to assure effective capacity use. Similar to the large technical system load problem described by Hughes (1987), large national and multinational corporations became interested in the full control of internal and external conditions affecting output and capacity use.

Besides serving these logistical functions, the new communications infrastructures also stimulated administrative rationalization. They improved coordination, monitoring and control from a distance. Through telegraphy, the central office of a geographically scattered firm could be informed instantly about new events and could just as quickly react appropriately. Railroad companies were the first organizations to use these new control technologies for internal administration and the centralization of control (Beniger 1986). Sola Pool (1983) made similar observations regarding the use of telephony. The telephone required much less operational skill than the telegraph and could, therefore, diffuse much quicker into the internal communications structure of a firm. Management from a distance and central control became much more "real time" than before.[8] Both telegraphy and telephony can therefore be considered important coordination technologies enabling a degree of control over production and distribution processes transcending vast geographical distances that was inconceivable before.

Telephony had some influence on the geographical decentralization and internal differentiation of corporations. It encouraged the separation between office and plant and thus supported the emergence of managerial hierarchies. While company offices were located at the production plant in the mid-nineteenth century (Hymer's second floor!), by the late nineteenth and early twentieth century, most corporate headquarters were separated from factories and concentrated in business districts such as Manhattan and Chicago. This phenomenon rationalized administrative activities and generated scales of economy in overhead functions as well, which is one of the major internalization advantages of large-scale hierarchies as conceptualized in transaction cost economics (Williamson 1985).

Current Globalization and the
New Communications Revolution

During the last decade, transnational business activities have been increasingly discussed under the heading of globalization. However, this globalization process was not a linear continuation of the internationalization take-off in the late nineteenth century. In some sense it was a reglobalization after the decline of world trade during the interwar period. Between the 1920s and the Second World War, waves of nationalism, growing tariff barriers and spreading webs of cartels led to a significant drag in the globalization process. Most of the multinational industries moved towards multi-domestic patterns. MNCs then mainly became federations of largely autonomous national subsidiaries (Porter 1987).

A new growth wave in foreign direct investment took off in the 1950s and accelerated in the 1970s. The earlier phase of this new internalization boom was mainly driven by American corporations. Since the mid-1970s, European multinationals have been very involved in the globalization process; somewhat later, the Japanese joined in. In the 1980s, the U.S. lost its hegemoneous position; in this new phase, the traditional pattern of the vertically integrated firm developed into divisionalized forms of more loosely coupled integration. The transformation was strongly influenced by strategies of product differentiation (Chandler 1977).

Quite contrary to the predictions of earlier market power theories, the new drive to globalization turned into intensive global competition. Increasing market pressure induced MNCs to exploit the division of labor among more geographically dispersed activities. Current internationalization processes imply not only backward and forward expansion, but the globalization of all elements of a firm's value chain; even R&D is increasingly global in structure (Porter 1987; Howells 1990).

The Communications Revolution

The new form of organizational differentiation and global integration would be unthinkable without the communications revolution in the 1950s and 1960s, which generated international telephony, commercial jet travel, satellite communications, and international computer networks (Chandler 1986). The coincidence between multinational growth and the boom in air traffic is amazing (Barnet and Müller 1974: 33). During the 1960s and 1970s, the air traffic system grew dramatically not only in scope but also in safety and reliability: "It affords safe passage at any hour, in almost any weather — usually to any airman who is qualified to seek it. It is a system that spans the globe, and reaches heights where the curve of the earth is visible" (La Porte 1988: 217). Increasing reliability and scope of air traffic fostered business globalization by serving as a speedy and reliable logistical link in the transportation chain — the dramatic growth of air freight is an indicator of this development — and by enabling a tighter control over foreign business through higher personal contact density (Chandler 1986: 152).

The most important forms of international business communications in the 1960s and 1970s were the telephone and the teletypewriter, also known as the telex. Although telephony had existed since the late nineteenth century and telex since the 1930s, the international diffusion of both systems did not become significant until the postwar period. Growth of international telephony was stimulated by the transatlantic telephone cables in the 1950s and even more by satellite technology in the 1960s. Both technologies increased the speed and reliability of information exchange.

The role of the telex in telecommunications development is often heavily underestimated. Until the 1980s, however, it was one of the most important communications instruments in international business (Antonelli 1985). Even in the most advanced industrial countries, the growth of the telex did not reach its peak until the mid- to late 1980s.

Only very recently has the telex been replaced by computer networks and other types of corporate telecommunications networks. One of the first international computer links was set up via telex lines in 1963 when Honeywell used a telex terminal in Britain to control a computer in a plant in Massachusetts (Chandler 1986:152).

The growth of computer networks during the 1970s was mainly based on leased telex and telegraph lines. During the same period, some global firms established their own private telecommunication systems. They were originally based on leased circuits from national telecommunications operators, but now such private networks can involve privately-owned satellite systems.

Multinational corporations interested in creating their own world communication systems were important social forces in the international liberalization of telecommunications (Schiller 1982). MCI's set-up of private microwave links in the 1970s is generally seen as key in eroding the traditional telecommunications monopoly in the U.S. and leading to the break-up of the Bell system (Galambos 1988). After liberalization in the United States, similar changes occurred in many industrialized countries (Dang N'guyen 1985, Schneider 1991), abolishing the major constraints for establishing private communications systems. Since then, corporate networks have grown dramatically (Irwin and Niman 1989).

As mentioned above, corporate telecommunication networks have to be conceived as "global managerial control technologies" facilitating current transformations in economic globalization. Corporate networks make the orchestration and synchronization of a firm's activities possible on a global scale: development, design, manufacturing and distribution may be internationally scattered but integrated into one single control structure. By interfacing with global transportation systems, production processes of MNCs become "geographically blind" (Irwin and Merenda 1989) and the global firm may operate in a global village.[9] A watch, car, computer and even a suite may be assembled by modules manufactured in different places in the world, interfaced by global transport logistics and coordinated via telecommunications networks. The combination of globally scattered production systems and infrastructures creates the "global industrial system constellations" Perlmutter (1972) predicted in the early 70s.

Global Competition and Information
Technology-Induced Reconfiguration

The current form of global division of labor by MNCs was a hot issue in academic debates in the 1970s. Some authors feared the emergence of an asymmetrical division of labor between industrial and developing countries in which labor-intensive production would be shifted to cheap labor areas in the world. Undoubtedly, one of the few location advantages of developing countries for industrial production is low wages; in some cases few or non-existing social and environmental regulatory standards can also be convenient.[10] As important as these location advantages are, they alone do not define a radically new division of labor in the world economy.[11] Nor is the retransfer of production processes from developing countries into automated factories of the developed world proof of a general trend in the other direction.

It is more plausible to assume a general tendency toward "global sourcing" with respect to a multiple set of location criteria. Products are manufactured completely or partly in different countries depending on location criteria specific to each sequence in the logistical or value chain of a firm. A basic premise is that components of finished products are bought internationally under the best conditions according to the slogan "we sell our products worldwide, therefore we must begin to buy and produce worldwide." But this form of global selling and sourcing is only possible on the basis of global LTS, such as telecommunications, air traffic, and sea and land transportation chains that integrate scattered manufacturing systems to global industrial production complexes.

The current phase of internationalization has often been labelled "global competition" to indicate that the global arena is a competitive system in which non-competitive firms will be driven out of the market. Global competition generates spatial homogenization in the sense that even local markets immediately become world markets, leaving no more room for a strategy of progressive expansion in one country after another. MNCs tend to use the same techniques for production and capital mobilization as well as the same organizational structures (Michalet 1991: 81-82).

Global competition exerts a whole series of pressures to increase in size and scope, to reduce the time to market, to provide better service, higher quality and lower costs, and to be more responsive to market changes. The major strategic responses to these pressures are outlined in Figure 1. They are technological sophistication (more R&D), shorter development-to-market times (simultaneous engineering and just-in-time production), higher flexibility and responsiveness (organizational decentralization, more customized products, and supplier out-sourcing.) Simultaneously, corporations try to improve quality by concentrating on fewer and more stable supplier relationships (single sourcing, supplier networks).[12]

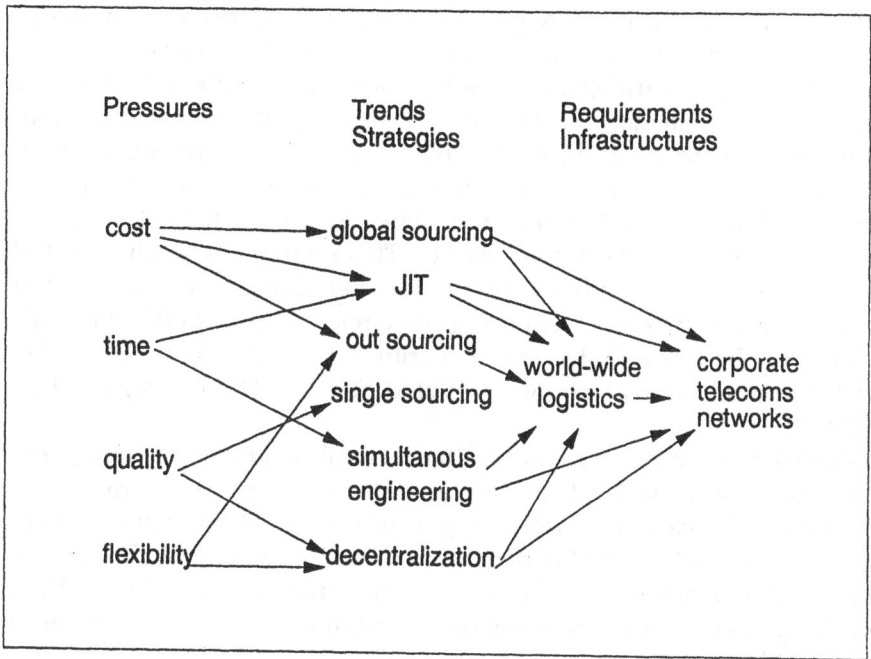

FIGURE 1   Global competitive pressures and corporations' strategic responses.

The combination of these competitive strategies leads to significant changes in the organization within and between firms, and within whole industry structures. The traditional organization of production during the Ford era (mass production of standardized products, strict division of labor and predominantly single-site manufacturing) transmutes into structures that are — paradoxically — more globalized, decentralized and technically integrated than before. This is certainly not the end of mass production, but mass production and distribution have become more flexible than in the Ford period (Sauer 1992).

Some of these changes were made possible largely through innovations in communications and information technology. New technologies allow greater decentralized operation at more widely dispersed locations, while simultaneously enabling higher and more effective integration and much tighter technical coupling within and across a production complex. A well-known example of tight technical integration is "simultaneous engineering" in design and manufacturing processes based on CAD/CAM via international computer networks.

Another case of increased global integration is international financial management through computer networks, in which cash flows and currency and borrowing are globally managed and respond in almost real-time to interest rates and currency exposures (Hagström 1992).

A further structural effect of global competition is the emergence of "network" or "cluster" firms through intra-organizational decentralization, out-sourcing and single sourcing strategies. Many multinationals have decentralized their organization to get greater responsiveness and flexibility. At the same time, however, they have to become larger to survive in the world market. This requires increasing central control and decentralizing action autonomy simultaneously (Rockart and Short 1991)[13] — i.e. "squaring the circle." These conflicting goals could only be reconciliated through critical innovations in information and communications technology (Venkatraman 1991, Applegate et al. 1988).[14]

Global computer networks facilitate central monitoring of company-wide activities, even when they operate in a highly decentralized, even globally scattered context (e.g. profit center organization, strategic business units).[15] Via corporate communications networks, central and local managers have instant and transparent control over their whole system of factories, warehouses and offices — in every corner of the world.

A sub-stratagem of decentralization is the reduction of a company's internalized activities by outsourcing. This strategy is also greatly supported by information technology. Suppliers may be integrated into specialized communications networks, allowing formatted data exchange (EDI).[16] Large supplier networks even enable forms of electronic subcontracting and electronic markets.

It is obvious that for globalized and decentralized firm structures, the development of logistical and communicative support systems becomes absolutely vital (Paché 1988). Increasing technical integration and the tighter interworking of these support systems transforms MNCs into globally nested LTS with intensities of coupling comparable to "classical LTS". Within the so-called second-order LTS (see Braun and Joerges, this volume), corporate telecommunications networks become a kind of nervous system linking and integrating the local production systems within a given firm's network.

In today's phase of globalization, a company's ability to compete depends heavily upon the sophistication of its corporate telecommunication network.[17] Most of the largest multinationals have been aware of the strategic implications of this fact for many years (Wiseman 1985).

One of the first strategic users of corporate networks was General Motors, which operated a company network based on private networks and leased carrier facilities in some fifty countries. The network provides Electronic Data Interchange (EDI) interfaces with 10,000 dealers and 20,000 suppliers (Irwin and Niman 1989).

General Electric, too, established its private network very early on. For many years, GE has offered its network facilities to third parties. GE's network provider GEIS offers information exchange and transactions facilities, financial transactions, and flight and hotel reservations. International freight companies coordinate container pools and freight volume through this technical infrastructure; the network enables a global positioning of each cargo around the globe.

Ford Europe recently installed a corporate network based on ISDN, linking Ford suppliers and customers in Europe and other continents via some sixty data centers. Ford sees the network as a major resource to support its strategies to shorten development-to-market time, improve quality and decrease costs. In R&D, Ford introduced international CAD/CAM networks based on advanced communications technologies for simultaneous engineering.

IBM was one of the first companies to create a private satellite communications system. Currently IBM also uses its corporate network to coordinate R&D activities worldwide.[18]

Hewlett-Packard is another strategic user of corporate communications. Its more than fifty manufacturing plants worldwide are linked by a private communications network. One of its major applications is a centralized procurement system.

Alcatel established a corporate network connecting 150 cities, 25 countries and 50 access points; about 25 suppliers were using EDI with Alcatel in 1991.

Volkswagen has factories in 20 countries on four continents. This global production system is supported by a highly centralized corporate network integrating almost all elements of the firm's value chain. VW operates international CAD/CAM systems, supplier integration by Electronic Data Interchange, and just-in-time systems with integrated transport logistics. A further application is a central system for technical information.

Siemens has developed its own corporate network, integrating about 200 production locations around the world (for more details, see Schneider and Ollmann 1991). One interesting application is logistical integration, as two logistics managers at Siemens write:

Worldwide logistical chains require integrated and speedy electronic information flows with the highest transmission quality. Therefore, Siemens is pushing the application of Electronic Data Interchange (EDI) within the intra-unit and cross-firms business. The decision to introduce the Edifact standard for all internal and external communications flows provides customers, suppliers and service providers with the necessary security for future-oriented investment in open communications systems. The use of standardized "barcode" technologies for the identification of freight units in the logistical chain "supplier — transportation — customer" links the commodity flow with a commodity preceding information flow. The expansion of the Siemens Corporate Network supports this global electronic capacity of communications among all Siemens units. (translated by Volker Schneider)[19]

Conclusion:
Toward the Post-Chandlerian Firm Complex

The preceding sections have shown that recent innovations in communication technologies facilitated fundamental transformations in production and industry structures. Multinational corporations are looking for increasingly decentralized corporate structures. Many are downsizing their hierarchical roofs to cover only core competencies while shifting peripheral activities to outside suppliers. At the same time, however, the suppliers' production systems become more and more tightly coupled and increasingly synchronized with the core corporation's production system, resulting in linking whole clusters of firms together. Industrial production systems of global scale emerge.

The technical coupling of production systems among separate firms reaches intensities which, in the Marshallian and the Chandlerian firms, could only be achieved within one factory or one corporate hierarchy. In the Marshallian firm, the boundary of the firm's production system was, as a rule, congruent with the boundary of the corporation. In the Chandlerian firm, the boundary of hierarchy became expanded via horizontal and vertical integration. Multiple production systems and even large technical infrastructures became integrated in ever-growing hierachies.

In the post-Chandlerian firm complex, however, this trend seems to be reversed: now it is the technical production system including large technical infrastructures which transcend the boundaries of individual firms, linking clusters of firms technically together (see Figure 2).

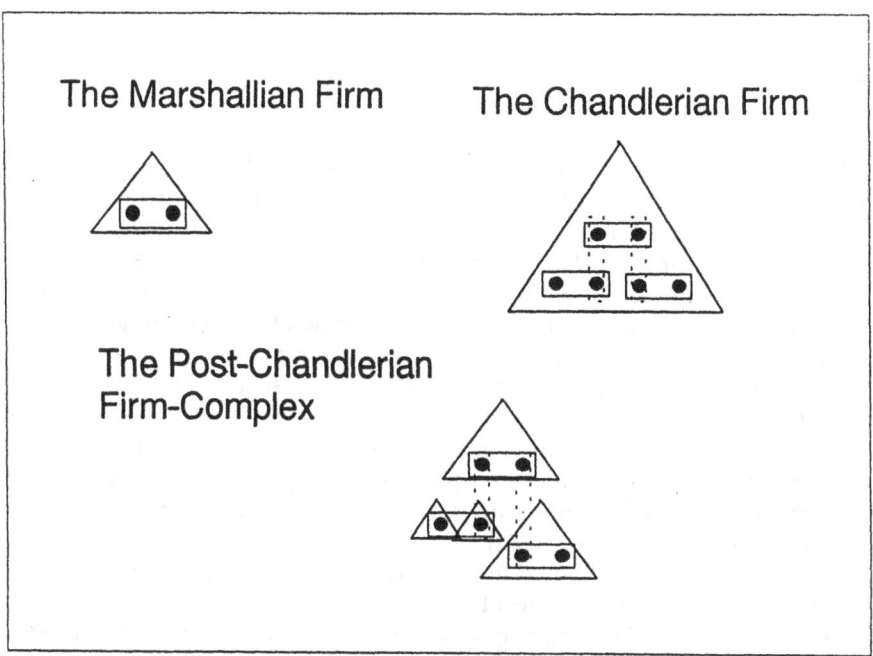

FIGURE 2 Coupling of technical production systems in Marshallian, Chandlerian, and post-Chandlerian firms.

Communications and transportation systems enable the logistical integration of globally scattered production units. Global production constellations become increasingly similar to classical LTS in which a number of different organizations may fulfill different sociotechnical roles in an overarching global technical complex.

Acknowledgements
I benefitted from comments by Lars Ingelstam, Karsten Ronit and Susanne Schmidt. Thanks to Cynthia Lehmann for linguistic assistance.

Notes

1. This is not only done for relatively simple processes based on cheap labor but also for more qualified activities like software programming. Texas In-

struments, for instance, imports software from India by satellite and American Airlines imports remote entry data from Barbados and the Dominican Republic (Irwin and Niman 1989).

2. It was already well known in classical social theory. Adam Smith, for instance, saw important stimuli for the division of labor and the growth of markets (see especially the third chapter in *The Wealth of Nations*) in the development of transportation. In Durkheim's *De la Division du Travail Social*, the growth of communications infrastructures ranks among the three major causes of functional differentiation: "Enfin, il y a le nombre et la rapi les vides qui séparent les segments des voies de communication et de transmission. En supprimant ou en diminuants sociaux, elles accroissent la densité de la société." (Durkheim 1986: 241)

3. Early corporations within the predominantly agrarian world of the Middle Ages in Europe were towns, guilds, hospitals and colleges. For an institutional derivation and discussion of medieval mediterranean merchant societies from Roman institutions, see Weber (1889).

4. Paper manufacturing, known in China for 2,000 years, did not spread to continental Europe until the fourteenth century.

5. Double-entry bookkeeping began with the development of the commercial republics of Italy; the first instruction manuals emerged during the fifteenth century.

6. In 1500, the first continuous postal line was introduced between Brussels and Vienna by the noble house of Thurn and Taxis, the first private postal operator in western and central Europe.

7. These fundamental changes in the transmission of commodities and information in the nineteenth century were characterized by Albion (1932) as the "communication revolution" as distinguished from the "industrial revolution." Similar observations were made by Borchardt (1973: 143), who characterized the nineteenth century as the "century of the revolution in communications."

8. The telephone, in fact, supported centralization: it improved hierarchical control and reduced the authority of field office managers considerably. An article in *Telephony* (1906) made the following observation: "[The Telephone]...has curtailed the functions and responsibilities of a district manager as the cable has those of an ambassador. "(Sola Pool 1983).

9. For this notion, see Bornschier (1988: 136) who characterizes the current phase of international corporate growth as "Weltkonzerne, die die Welt mit ihrer neuen Informationstechnik zum Dorf machen."

10. This issue is discussed in Schneider (1987).

11. For the new division of labor idea, see Fröbel et al. (1977) and Hymer (1972). A critique of this approach is given in Ernst (1986) and Schoenberger (1989).

12. With respect to the megatrends "global sourcing, out-sourcing and single sourcing" see Faulhaber and Schmitt (1992: 139-144).

13. A similar observation was made by Klebe and Roth (1987: 26) with regard to the strengthening of centralized control by computer networks.

14. See the conceptualization of this change by Child (1987) on the basis of transaction cost theory. For Child, the "network firm" corresponds to the development of coordinated contracting arrangements by previously integrated firms enabled through new information technology. A key managerial requirement is to apply the same degree of control and coordination to transactions within a network of firms as could be applied within one integrated firm.

15. An example is Xerox's recently established executive support system. This system requires all of Xerox's thirty-four business units to submit their plans via the company's communications network in a specific format. Each business unit's plan can then be reviewed not only by senior executives and corporate staff but also by other top officers in the firm. Another example is Philips Petroleum Company's executive support system, linking the ten top executives to one another and providing on-line access to varying levels of daily sales, refinery, and financial data (Rockart and Short 1991).

16. See Aigner and Kuckelkorn's (1991) observations about Ford. Since the 1980s, Ford has reduced the number of its suppliers by fifty percent. Its explicit goal is to have only one supplier per region. Ford aims to connect all suppliers to a global electronic system of material purchasing control.

17. Corporate information and communications systems are being discovered as a distinct element of competitive strategy (Wiseman 1986). Important goals are to cut transaction costs, accelerate product design, quicken product introduction, coordinate scattered manufacturing operations and interface with suppliers by Electronic Data Interchange (Irwin and Merenda 1989; Irwin and Niman 1989).

18. An example of the use of a corporate network for coordinating the development of IBM's latest version of the 9,370 mainframe computers is given by Guterl (1989: 33). In this case the "participants were linked by electronic mail, teleconferencing and other communications channels and they were forged into working relationships through ad hoc project teams, multidisciplinary conferences and the like."

19. Blick durch die Wirtschaft, 1 April 1992.

References

Aigner, Jürgen and Wilfried Kuckelkorn. 1991. Die weltweite Verflechtung konzerneigener und selbständiger Lieferbetriebe im Hause Ford. In *Zulieferer im Netz — zwischen Abhängigkeit und Partnerschaft: Neustrukturierung der Logistik am Beispiel der Automobilzulieferung*, ed. Hans G. Mendius and U. Wendeling-Schröder, 131-142. Köln: Bund-Verlag.

Albion, R. G. 1932. "The Communication Revolution." In *American Historical Review*, 37: 718-720.

Antonelli, Cristiano. 1985. "The diffusion of an organizational innovation. International Data Telecommunications and Multinational Industrial Firms." In *International Journal of Industrial Organisation*, 3:109-118.

Applegate, Lynda M., James I. Cash and D. Quinn Mills. 1988. "Information Technology and Tomorrow's Manager." *Harvard Business Review* (Nov.-Dec. 1988).

Arnold, Ulli. 1990. "Global Sourcing" — Ein Konzept zur Neuorientierung des Supply Management von Unternehmen. In *Globales Management: Erfolgreiche Strategien für den Weltmarkt*, ed. Martin K. Welge, 49-71. Stuttgart: Poeschel.

Barnet, Richard J. and Müller, R. E. 1974. *Global Reach. The Power of the Multinational Corporations*. New York: Simon & Schuster.

Beniger, James R. 1986. *The Control Revolution. Technological and Economic Origins of the Information Society*. Cambridge, Mass.: Harvard University Press.

Borchardt, Knut. 1973. The Industrial Revolution in Germany 1700-1914. In *The Emergence of Industrial Societies. The Fontana Economic History of Europe*, Part One, ed. Carlo M. Cipolla, 76-90. Collins: Fontana.

Bornschier, Volker. 1988. *Westliche Gesellschaft im Wandel*. Frankfurt: Campus.

Casson, Mark. 1990. *Multinational Corporations*. Aldershot: Elgar.

Chandler, Alfred D. 1977. *The Visible Hand. The Managerial Revolution in American Business*. Cambridge, Mass: Harvard University Press.

_____. 1980. "The Growth of the Transnational Industrial Firm in the United States and the United Kingdom: A Comparative Analysis." *The Economic History Review* 33: 396-410.

_____. 1984. "The Emergence of Managerial Capitalism." *Agricultural History* 58(3): 473-503.

_____. 1986. "Managers, Families, and Financiers." In *Development of Managerial Enterprise. The International Conference on Business History 12. Proceedings of the Fuji Conference*, ed. K. Kobayashi, H. Morikawa, 35-63. Tokyo: University of Tokyo Press 1986.

_____. 1989. Die Entwicklung des zeitgenössischen globalen Wettbewerbs. In *Globaler Wettbewerb. Strategien der neuen Internationalisierung*, ed. M. E. Porter , 467-514. Wiesbaden: Gabler.

_____. 1990. *Scale and Scope. The Dynamics of Industrial Capitalism*. Cambridge, Mass.: Belknap Press.

Child, John. 1987. "Information Technology, Organization, and the Response to Strategic Challenges." *California Management Review* 30.

Dang Nguyen, Godefroy. 1985. "Telecommunications: A Challenge to the Old Order." In *Europe and the New Technologies*, ed. Margaret Sharp, 87-133. London: Pinter.

Durkheim, Emile. 1986. *De la Division de la Travail Sociale*. Paris: Presse Universitaire de France.

Ells, Richard. 1972. *Global Corporations. The Emerging System of World Economic Power*. New York: Interbook.

Ernst, Dieter. 1986. "Innovationen in der Mikroelektronik und die Internationalisierung der Elektroindustrie." In *Technik und internationale Politik*, ed. Beate Kohler-Koch, 89-138. Baden-Baden: Nomos.

Faulhaber, Peter and Helmut Schmitt, 1992. "Die Mega-Trends der Kraftfahrzeug-Zulieferindustrie." In Europa 1992. *Strategie — Struktur — Ressourcen,* ed. Heinz W. Adams, 139-155. Frankfurt: FAZ.

Fröbel, Frieder, Jürgen Heinrich and Otto Kreye. 1977. *Die neue internationale Arbeitsteilung. Strukturelle Arbeitslosigkeit in den Industrieländern und die Industrialisierung der Entwicklungsländer.* Reinbek: Rowohlt

Galambos, Louis. 1988. Looking for the Boundaries of Technological Determinism: A Brief History of the U.S. Telephone System. In *The Development of Large Technical Systems,* ed. Renate Mayntz, Thomas P. Hughes, 135-154. Frankfurt: Campus.

Guterl, Fred V. 1989. Goodbye, Old Matrix. *Business Month* 133: 32-38.

Hack, Lothar and Irmgard Hack. 1990. Gestaltung — Erzeugung — Erbauung. Industrieforschung als strategische Einrichtung zur Produktion sozialer Realität. In *1990 — Jahrbuch Arbeit und Technik,* ed. Werner Fricke, 243-256. Bonn: Dietz.

Hagström, Peter. 1992. "Inside the 'Wired' MNC." In *The Economics of Information Networks,* ed. Cristiano Antonelli , 325-340. Amsterdam: North-Holland.

Hepworth, Mark E. 1989. *Geography of the Information Economy.* London: Belhaven Press.

Howells, Jeremy. 1990. "The Globalization of Research and Development: A New Era of Change?" *Science and Public Policy* 17(4): 273-285.

Hughes, Thomas P. 1987. "The Evolution of Large Technical Systems." In *The Social Construction of Technological Systems,* ed. Wiebe Bijker, Thomas P. Hughes, Trevor J. Pinch, 51-82. Cambridge, Mass: MIT Press.

Hymer, Stephen. 1972. "The Multinational Corporation and the Law of Uneven Development." In *Economics and World Order,* ed. Jagdish N. Bhagwati, 113-140. London: Macmillan.

Irwin, Manley R. 1989. "Corporate Networks, privatization and state sovereignty. Pending issues for the 1990s?" *Telecommunications Policy,* 13(4): 329-335.

Irwin, Manley R. and Neil B. Niman. 1989. "The Corporate Telecommunications Network: Market Transparency and State Accountability." In *European Telecommunications Policy Research,* ed. Nicholas Garnham, 175-182. Amsterdam: IOS.

Klebe, Thomas and Siegfried Roth. 1987. "Informationen ohne Grenzen. Globaler EDV-Einsatz und neue Machtstrukturen." In *Informationen ohne Grenzen. Computernetze und internationale Arbeitsteilung,* ed. Thomas Klebe, Siegfried Roth, 7-42. Hamburg: VSA-Verlag.

La Porte, Todd R. 1988. "The United States Air Traffic System: Increasing Reliability in the Midst of Rapid Growth." In *The Development of Large Technical Systems,* ed. Renate Mayntz and Thomas P. Hughes, 215-244. Frankfurt: Campus.

Mayntz, Renate. 1988. "Zur Entwicklung technischer Infrastruktursysteme." In *Differenzierung und Verselbständigung. Zur Entwicklung gesellschaft-*

licher Teilsysteme, ed. R. Mayntz, B. Rosewitz, U. Schimank, R. Stichweh, 233-259. Frankfurt: Campus.

Michalet, Charles-A. 1991. "Global Competition and its Implication for Firms." In *Technology and Productivity.* The Challenge for Economic Policy. Paris: OECD.

Miles, Raymond E. and Charles C. Snow. 1986. "Organizations: New Concepts for New Forms." *California Management Review* 28(3): 62-73.

Mulgan, Geoff J. 1991. *Communication and Control. Networks and the New Economies of Communication.* Cambridge: Polity Press.

OECD. 1992. *Globalization of Industrial Activities. Four Case Studies: Auto Parts, Chemicals, Construction and Semiconductors.* Paris: OECD.

Paché, Gilles. 1988. "La Prise en Compte des Techniques de Télécommunications dans l'Analyse des Réseaux Logistiques: Réflexions Méthodologiques." In *Information et Organization Spatiale,* ed. Henry Bakis, 93-102. Caen: Paradigme.

Parry, T. G. 1985. "Main Trends in the Evolution of Multinational Enterprise Structures." In *Transborder Data Flows,* ed. Hans-Peter Gassmann, 173-205. Amsterdam: North-Holland.

Perlmutter, Howard V. 1972. The Multinational Firm and the Future. *The Annals of the American Academy of Political and Social Science* 403: 139-152.

Peters, Tom. 1992. *Liberation Management. Necessary Disorganization for the Nanosecond Nineties.* London: Macmillan.

Piore, Michael J. and Charles F. Sabel. 1984. *The Second Industrial Divide. Possibilities for Prosperity.* New York: Basic Books.

Porter, Michael E. 1987. "Changing Patterns of International Competition." In *The Competitive Challenge. Strategies for Industrial Innovation and Renewal,* ed. David J. Teece , 27-57. New York: Harper & Row.

Rockart, John F. and James E. Short. 1991. "The Networked Organization and the Management of Interdependence." In *The Corporation of the 1990s. Information Technology and Organizational Transformation,* ed. Michael S. Scott Morton, 189-219. New York: Oxford University Press.

Sauer, Dieter. 1992. "Auf dem Weg in die flexible Massenproduktion." In *Vernetzte Produktion. Automobilzulieferer zwischen Kontrolle und Autonomie,* ed. Manfred Deiss and Volker Döhl , 49-79. Campus: Frankfurt.

Schneider, Volker. 1991. "The Governance of Large Technical Systems: The Case of Telecommunications." In *Responding to Large Technical Systems: Control or Anticipation,* ed. Todd R. La Porte, 18-40. Dordrecht: Kluwer, 1991.

_____. 1987. "Transnationale Chemikalienkontrolle: Internationale Technikentwicklung in einer Kontroll-Lücke?" In *Technikkontrolle und internationale Politik. Internationale Steuerung von Technologietransfers und ihre Folgen,* ed. U. Albrecht , 195-219. Opladen: Westdeutscher Verlag, 1989.

Schneider, Volker and Rainer Ollmann. 1991. "Transnational Use of High Speed Data Communications and Digital Video Services. Case Studies — Germany." In *Perspectives for advanced communications in Europe: 1990. Volume VII: IDS 4 — Transnational Applications in Europe.* Brussels:

Commission of the European Communities, DG XIII, Directorate F (RACE) C1-35.

Schoenberger, Erica. 1989. "Multinational Corporations and the New International Division of Labor: A Critical Appraisal." In *The Transformation of Work? Skill, Flexibility and the Labor Process*, ed. Stephen Wood, 91-359. London: Unwin Hyman.

Smith, Adam. [1789] 1974. *Der Wohlstand der Nationen*. München: DTV.

Sola Pool, Ithiel de. 1983. *Forecasting the Telephone: A Retrospective Technology Assessment*. Norwood: Ablex.

United Nations Centre on Transnational Corporations 1988. *Transnational Corporations in World Development. Trends and Prospects*. New York: United Nations.

Venkatraman, N. 1991. IT-induced Business Reconfiguration. In *The Corporation of the 1990s. Information Technology and Organizational Transformation*, ed. Michael S. Scott Morton. New York: Oxford University Press.

Weber, Max. [1889]1988. "Zur Geschichte der Handelsgesellschaften im Mittelalter." In *Gesammelte Aufsätze zur Sozial- und Wirtschaftsgeschichte*. Tübingen: Mohr.

Wieland, Bernhard. 1990. "Multinational Banking, the Theory of the Division of Labor and the Information and Communications Technologies." In *Communications Policy in Europe*. Proceedings of the 4th Annual Communications Policy Research Conference, ed. Dieter Elixmann and Karl Heinz Neumann. Berlin: Springer.

Williamson, Oliver E. 1985. *The Economic Institutions of Capitalism*. New York: Free Press.

Wiseman, Charles. 1985. *Strategy and Computers: Information Systems as Competitive Weapons*. Homewood: Dow Jones-Irwin.

5

Changing Embedded Systems: The Economics and Politics of Innovation in American Railroad Signaling, 1876-1914

Steven W. Usselman

Introduction

Over a span of approximately thirty years around the turn of the twentieth century, American railroads remade their systems for handling trains. At the start of that period, virtually all lines governed the movement of trains by the dispatching or time-interval system. In these schemes, dispatchers used telegraphic connections to monitor the movement of trains from station-to-station and issue orders to train crews about where and when to proceed. By prohibiting trains from departing too soon after one another and enforcing strict rules about speeds, a safe distance between trains could be maintained as long as crews of unexpectedly delayed trains provided ample warning of their presence on the track.

Thirty years later, most American railroads had come to rely on a fundamentally different method of train-handling. Known as the block system, this method called for railroads to maintain space between trains by dividing the track into sections (or blocks) and prohibiting any train from entering a section before the previous train had left it. If followed, this procedure could remove the need for dispatchers and enable railroads to simplify the rules of operation dramatically, while providing virtually fail-safe protection against collisions. "The block system is simple and scientific," the Interstate Commerce Commission (ICC) stated succinctly in its 1905 report, "while the 'time-interval' system is complicated and difficult."[1]

This glib assessment masks an extraordinary complexity that the conversion to the block system entailed. Far from the straight-forward simplification of procedures implied by the ICC, adoption of block signaling often posed challenging complications. The decision to adopt it involved a myriad of factors that few railroad managers saw in precisely the same way. Consequently, the block system did not sweep across the industry. Despite conspicuous, sustained efforts by the ICC and other agents of comprehensive reform, it spread in fits and starts, and American railroads — both individually and collectively — ended up with a hybrid system for governing the movements of trains. This chapter examines why this happened and suggests that the answer lies in the complex interplay among several systems.

Conceptual Overview

To trace the convoluted course of change in railroad signaling and explain its outcome, this chapter will draw heavily on two concepts used when discussing technological systems. First, signaling is an *embedded* system. It exists within a larger, more comprehensive system, that is, the railroad itself. Changes in the signaling system cannot be understood apart from that larger system. In the case of railroad signaling at the turn of the century, moreover, the larger system was itself embedded in two additional systems. Collectively, the individual railroad companies responsible for signaling made up a national railroad system, which through various forms of cooperation and merger was attaining a greater degree of integration. That process of consolidation was hastened by efforts to regulate railroads at the federal level, a task which formed a major component in the drive to create a system of national economic administration. In effect, change in railroad signaling involved a hierarchy of systems: the regulatory system, the national railroad system, the individual railroads and the signaling systems themselves.

The notion of embedded systems is in some respects a variant of the second concept, which is that most systems exhibit some degree of *indeterminancy*. Though all systems impose constraints that demand conformity with their constituent parts (without such qualities, the term "system" would hardly seem to possess meaning), most retain the freedom to accommodate some variation. For instance, all railroads consist of rails, locomotives, cars, stations, yards, and personnel but no two will deploy precisely the same mix of these components. When separate railroad systems are joined or embedded in a national system, that larger system will exhibit a variation of a sort that might best be labeled *differentiation*. It appears to consist of several identical components, but on closer inspection one can detect subtle differences between them.

The key to analyzing any system is to assess its level of indeterminancy. One must decipher the balances struck between conformity and freedom, determine what features are negotiable, and decide the degree of differentiation that can be tolerated within any embedded system. When examining systems in transition, the key issues become whether the changes fit into the indeterminancy of the system and whether one can tolerate any differentiation they might cause.

Characteristics of the Block System

The block system possessed considerable indeterminancy. While in theory it referred to a single concept of handling trains, in practice it appeared in many different guises. One important source of variation involved the use of rules that permitted engineers in certain circumstances to proceed into an occupied block. This practice, referred to as permissive blocking, might apply to certain types of trains or to certain hours of service. The most dramatic variations, however, involved how news that a train had entered or exited a block was conveyed from one end to the other and passed along to the crews of approaching trains. This *could* be done quite simply, and often was. Station agents, linked by telegraph, might serve as block monitors for the track between their stations and communicate with engineers by handing them a note or displaying a colored flag. But these functions also opened up vast possibilities for innovation in signaling, including the use of electrical apparatus that automatically detected the presence of a train and set the appropriate warnings without human input. Conversion to the block system sometimes entailed the construction of complex new electro-mechanical systems, as well as extensive revision of rules.

With its inherent flexibility, the block system meshed well with the needs of its host system, the railroads. Despite commonalities that enabled them to function in some respects as a single system, railroads were a highly differentiated lot. The extent of change associated with block signaling consequently varied dramatically from line to line, as managers confronted different and rapidly changing circumstances and weighed the benefits of technical options that were rapidly proliferating. Decisions on conversion involved many factors so complex that government investigators and other contemporary observers often despaired to make sense of them.[2]

In its most basic form, the block system offered potential benefits along three distinct but interrelated paths. First, it reduced the chance of collisions because space would be maintained even if trains broke down or did not travel at expected speeds. Second, it made railroads less dependent on craft procedures or "the human element" since the movement of trains would no longer require dispatchers capable of making active, impromptu decisions or crews that could be counted on to execute orders. Third, it provided an opportunity to run more trains through a given length of track because trains proceeding through a set of short blocks could follow one another more closely than the smallest permissible time interval allowed. Virtually every variation of the block system offered a different mix of these basic benefits (and carried its own cost burdens) and virtually every railroad manager placed a different value on them.

To a large extent, conversion to the block system must be studied as many separate episodes. One aim of this chapter is to trace one such case — that of the Pennsylvania Railroad — in considerable detail. Because it operated the most intensively used track system among major American railroads, the Pennsylvania took the lead in establishing block signal stations. During the last quarter of the nineteenth century, the company gained experience in running trains under the block system and grappled with its many implications for its labor force and methods of accounting. By 1900, the Pennsylvania had begun to automate its block system, and again it confronted various ramifications for its organization. During this sequence of alterations, managers at the line were concerned with adjusting their facilities to handle a massive increase in traffic in the most economical fashion. Signaling technology and the reconfigured system for handling trains contributed mightily to this effort. By revising its system for handling traffic, the Pennsylvania avoided more costly alterations to its physical plant. Reconfiguring the embedded system helped preserve the host.

It would be a mistake, however, to interpret the rise of the block system as merely the sum of decisions and activities at individual lines. Such a view would mask important dimensions of the change. For while a few companies may have acted as the Pennsylvania did and made the switch to the block system sheerly out of concern for internal operational considerations, the wholesale conversion of nearly 100,000 miles of track can ultimately only be understood as a collective response to political pressures that acted upon the railroad system as a whole. The impetus for change worked its way down the hierarchy of systems as well as up, and as it did so, the issue of railroad signaling was absorbed into a broader set of concerns pertaining to systems higher up the chain.

The ICC's Interest

By the early 1890s, reform-minded legislators had begun to create a regulatory apparatus capable of exerting considerable authority over railroad safety. Beginning in 1902, the ICC lobbied annually for a law requiring the block system, which the British Parliament had succeeded in implementing a decade earlier. Though it failed, the commission gained significant power to investigate and publicize, which it used to monitor progress in signaling and to campaign for the block system. ICC regulators also hoped to use the concerns about safety stimulated by discussion of the block system to gain additional supervisory authority over railroad operations. These actions pertaining directly to safety and the block system, moreover, occurred within a politically charged atmosphere that continually threatened railroads with unsolicited regulation regarding rates, wages, and other matters they held very dear. If railroad managers appeared overly callous or parsimonious on safety matters, they might damage their reputation and hurt their chances in more critical political battles. Just as the Pennsylvania sought to reconfigure its signaling system to avoid greater alterations to its plant, railroad companies tried collectively to respond to the agitation for block signaling in a way that would keep that agitation from spilling over into other, more important issues to the whole railroad system.

Perhaps the greatest challenge associated with this industry-wide effort — and certainly one of its most interesting elements for students of systems in transition — was that railroad officials wanted to reach a solution that could accommodate the peculiar characteristics of the individual lines that made up the system. They wanted to preserve a high degree of differentiation in the national railroad system. The ongoing revisions to the signaling system at the Pennsylvania in many respects made this task more difficult. At a time when few lines had considered even the simplest block system, the Pennsylvania's pioneering efforts drew widespread attention to sophisticated techniques developed to meet exceptional operating conditions. In the eyes of many, the automated techniques of the Pennsylvania offered an alternative to routine supervision of procedures by the ICC. If safety advocates had succeeded in requiring the Pennsylvania's captivating innovations, few other railroads could have borne the costs so easily.

In the end, however, regulation would not go so far. The reconfigured system of governing rail traffic instead relied on intensified publicity and the threat of legislation to encourage adoption of procedures that were in fact much more varied and flexible than the single name "block system" implied.

Innovation at the Pennsylvania: Imperatives of Congestion

In its broad outlines, the revision of methods governing train movements on the Pennsylvania Railroad followed a simple, logical course. By the last quarter of the nineteenth century, the Pennsylvania had built multiple-track routes through some of the United States' most economically active and densely populated areas. During the next four decades as these regions underwent extraordinarily rapid industrial development, the railroad continually struggled to move more goods through its network of tracks. This effort involved a classic exercise in systems management, in which the company's managers constantly juggled factors such as the size and speed of trains, the condition of tracks, the number of sidings, the hours of operation and the procedures for assembling trains.

Signaling changes and the rules governing train movements contributed to this effort by enabling the Pennsylvania to move more trains through a stretch of track during a given period. More frequent trains posed risks that necessitated such changes and at the same time generated revenue that justified larger expenditures on infrastructure, such as signaling. As traffic grew denser, the Pennsylvania turned to more capital-intensive methods of signaling. It first built manually-operated block stations at distant intervals, later added more stations as the need arose, and ultimately switched to automated techniques. Textbooks and governmental studies of the early twentieth century regularly traced this sequence and routinely held up the Pennsylvania as the exemplar of modern signaling.

While correct in its broad outlines, this summary conveys an image of an orderly transition that conceals the many complexities that managers at the Pennsylvania struggled with along the way. Documents from the firm's archives, recently opened to historians, suggest that managers grappled throughout the period to master rapidly changing technology and fit it into existing assumptions about how best to operate a railroad. In essence, the managers of the Pennsylvania continually had to redefine what the block system was and reevaluate what purposes it served. They struggled in particular to find ways to take advantage of the order and security new technologies could provide without sacrificing the flexibility necessary to meet varied and changing conditions.

Expansion of the Block System

The Pennsylvania made its first significant departure from the dispatching system during the mid-1870s, when it constructed block signal stations along its main route across New Jersey from New York to Philadelphia and laid plans to extend the system westward toward Harrisburg and Pittsburgh. These routes all had multiple tracks on which trains travelled in only one direction. Attendants in the towers, linked by telegraph, regulated the movement of trains by displaying a stop signal when a train passed their station and changing it to "go" when they knew the train had passed the next tower.[3] In constructing the network of stations dedicated exclusively to signaling and operating it in this fashion, the Pennsylvania jumped ahead of the rest of the industry by at least a decade. No other American railroad had a similar system prior to 1885.[4]

At about the same time it installed the block system, the Pennsylvania also began to purchase interlocking signals for its busiest crossings and switches. Interlockings linked signals and switches together in ways that guaranteed that all signals corresponded with the actual position of the track. When a switch was moved, the signals moved with it. These arrangements protected against derailments and insured that only one train could receive a "through" signal. Simple, mechanical interlockings powered by a switchman pulling a lever had become quite common in England during the 1870s. The Pennsylvania acquired its equipment from the leading British supplier, Saxby and Farmer.[5]

The few existing records from this period unfortunately provide little information about why the company took these pioneering steps. Increasing congestion obviously provided the general stimulus since the Pennsylvania limited the new methods to its busiest lines, but we do not know for certain what in particular concerned its managers or what they actually hoped to achieve. One document from a later period, however, offers some tantalizing hints. In an 1893 report intended to "form the foundation of the art of signaling on the Pennsylvania Railroad," a committee of managers stressed that it was offering "a radical change in regard to the function of signals." It called for the Pennsylvania to abandon its "unwritten axiom...that 'signals govern tracks'" and replace it with a frank admission that "signals are primarily for the purpose of controlling or indicating the movement of *trains*." In making this recommendation, the committee argued that the shift in approach had already begun. "There is certainly a tendency in handling the traffic on busy lines," it observed, "to make the signals a quick and positive means of conveying information or orders to trains."[6]

This fascinating document suggests that during the 1890s the Pennsylvania experienced a revolutionary change in attitude about the role of signals and the block system in its operations. Apparently managers initially viewed the system primarily as a means of providing additional safety and protection under demanding conditions. They used the system — just as safety advocates would later conceive it — to prevent more than one train from occupying the same stretch of track. By 1893, however, managers were coming to see signals, including those associated with the block system, as a convenient mechanism for communicating with trains in motion. In their eyes the block system was not so much a means of checking trains in crowded conditions but a tool for keeping trains moving and relieving congestion. Speed and flow were now the paramount objectives and the function of signaling was "to pass the maximum traffic over the road at the greatest speed."[7]

This shift in attitude apparently grew out of practical problems that built up under the old block system. As the number of trains increased and more blocks filled up, trains began to encounter stop signals with intolerable frequency. In 1894, members of the signal committee identified "the question of reducing the number of red signals displayed to an approaching train" as "one of two subjects that have been brought regularly before [us]."[8] Pressure mounted to clear blocks more rapidly by running at faster speeds, but block signals placed limits on speed because engineers would overrun them if traveling too fast. The Pennsylvania had dealt cleverly with this situation by adding an additional semaphore blade at each tower which provided engineers with information about the status of the block beyond the one they were about to enter. These "distant" or "caution" signals kept engineers from having to travel in anticipation of stopping at signals that might in fact be set in their favor. It was no doubt these caution signals, together with interlockings, that led to the realization that signals had become a means of "conveying information or orders to trains."[9]

Crisis Point

By the early 1890s, the old system of manually operated towers had reached a point of crisis even with the addition of distant signals. As the 1894 report indicated, stop signals continued to disrupt operations. In response, management had to some degree allowed the system to operate permissively. After stating emphatically that "the permissive block signal cannot be defended under any perfect signal system," the authors of the 1893 report acknowledged "the necessity under present conditions for a permissive block system."[10] A report issued in 1896, moreover, documented an alarming number of dangerous mistakes resulting from the manual operation of distant signals. It stressed the need to write additional rules for permissive blocking with distant signals.[11]

But new rules and permissive operations were clearly stopgap measures and all of these reports pointed to new technology as the ultimate solution. "The decreased cost and improvement of automatic block signals will in all probability in the near future make possible the absolute block system," wrote the committee in 1893, "where the question of expense now makes it practically impossible of adoption."[12] The second major subject considered by the committee in 1894 was "reducing the cost of interlocking, so that it may be put in more general use as soon as practicable."[13] Interlockings could be used to connect home and distant signals and they would enable engineers to proceed across switches and other dangerous points without slacking speed. Two years later the committee, considering new rules, suggested that perhaps the Pennsylvania should link home and distant signals by track circuit. "There are many other appliances which could be provided which would make signaling more safe," the committee went on to note in an apparent reference to the automatic block, "but as these would involve a considerable outlay of money, it seems important that the necessary rules should be made and enforced."[14]

By 1896, the Pennsylvania Railroad had come to the brink of radical reconfiguration. Its managers had grasped the fundamental limitations of the initial block system, reconceptualized the role of signals, identified alternatives, and created forums for discussing and implementing change. Apparently only the shortage of capital in the depressed economy of the mid-1890s held them back. However, sometime during the next two years the Pennsylvania began to make the switch to automatic signals.[15] By 1901, it had converted over 500 miles of line.[16]

Installation of the Automatic Signaling System

Unfortunately, available records do not provide much insight into the events that accompanied actual construction of the new automatic block system. From a distance we can identify certain features of the conversion that might have some significance for the broader subject of systems in transition.

Though the Pennsylvania's automatic signaling system unquestionably represented the pioneering, "state-of-the-art" facility of its kind, its installation does not appear to have posed many technical difficulties or required a long "shake-out" period. One major reason for this is that the Pennsylvania could rely on outside suppliers for the technology. George Westinghouse's Union Switch and Signal Company handled the task on a contract basis. Firms such as Westinghouse's had already established themselves as suppliers of signaling technology by designing and constructing customized interlocking plants at complex switching points. Many of the innovations used in automatic block systems originated with these interlockings. By the late 1890s, moreover, Westinghouse could draw on his extensive experience of installing electrical plants. These experiences gave Westinghouse a ready claim on authority that most suppliers to the railroads did not enjoy. Unlike locomotive builders and rail makers, he did not have to deal with an entrenched group of experts within the Pennsylvania.[17] In an odd way, conversion to the automatic block system was probably made easier by the fact that the Pennsylvania had little established expertise in the methods involved.

Freed from responsibility for design and invention, the Pennsylvania concentrated within its own staff on building up teams of inspectors, who were supervised by an executive eventually known as the signal engineer. Like other staff officers, the signal engineer had to cooperate with line officers responsible for operations and he worked closely with the maintenance of way department. As in the early 1890s, major policy decisions regarding signals were often made by committees that combined these perspectives. Textbooks on signal engineering claimed that on some roads these organizational issues created difficulties.[18] The records of the Pennsylvania contain a few examples of reports that languished in committee, most notably the "Rudd and Rhea Report," an attempt by a maintenance of way officer and a signal engineer to standardize signal blade positions and light colors. It stirred controversy within the Pennsylvania between 1905 and 1908 and became notorious throughout the industry.[19] On the whole, however, change seems to have been facilitated by the fact that the Pennsylvania already had established organizational mechanisms such as the Association of Transportation Officers that promoted intrafirm cooperation.

In assembling this expertise, the Pennsylvania also tapped groups that rapidly formed within existing railway associations to consider the subject of signals. As early as 1894, the signals committee attended a meeting of one such group at the American Railroad Association.[20] By 1905 a separate Railway Signaling Association was organized. The technical literature also helped diffuse information rapidly. Braman B. Adams monitored developments closely for the trade journal *Railroad Gazette* and in 1901 published a book on the subject of block signals. When a college textbook on signaling appeared in 1909, the science of signal engineering had clearly emerged.[21]

Implications for Labor

The creation of this force of experts and inspectors relates to the more general issue of how labor was affected by the switch from manual to automatic methods. Clearly, the Pennsylvania did not make the radical conversion it did to replace workers with machines. It made the change because the problems of congestion had simply overrun the ability of the existing track facilities to handle traffic using manually operated signaling methods. Electrical technology had not merely provided an opportunity to replace labor; it had opened up new possibilities.

While the substitution of machines for men had not motivated the change, the switch to automatic techniques did have profound implications for work and workers. The telegraph operators who had manned the old towers obviously felt some effects, though there were few of these men and it is not clear whether or not the Pennsylvania continued to use them in freight operations.[22] Regardless of the immediate effects at the Pennsylvania, its efforts to handle more traffic had led to developing techniques that might subsequently be used to reduce labor costs. Studies of signaling regularly compared the costs of automatic and manual methods, often using data from the Pennsylvania.[23]

Automatic signaling also had important effects on train crews. The new system took over responsibility for governing the speed of traffic and made engineers into drivers who raced from signal-to-signal, clearing blocks as rapidly as possible. This mode of operation furthered the ongoing transformation of engineers from individuals who were in charge of particular trains and were compensated by the route into hired operators paid by the hour. But again, this feature was an implication of the change rather than a motivation for it. Conversion to the block system did not immediately involve significant alterations in compensation or work assignments and engineers do not appear to have objected to it.

The automatic system did, however, raise a troubling issue of accountability. Under manual operations, managers had relied on engineers and signalmen to provide a mutual check against error. Now an engineer could pass a stop signal without being observed and blame mechanical malfunction of the signal if something went wrong as a result. The Pennsylvania resorted to surprise tests and disciplinary hearings to protect against this possibility. These procedures continually stirred controversy. When managers reviewed them in 1908 following publication of a series of essays about railroad discipline in the *Atlantic Monthly,* they recommended that the tests be renamed "efficiency tests." "The term 'efficiency' seems more dignified and satisfactory than the word 'surprise,' they noted, "and should give less offense."[24]

This discussion of employee discipline together with the extended debate over signal blades and light colors indicates that the reconfiguration of the automatic block system had run its course at the Pennsylvania. The new system had reached maturity. Any further significance it held for a study of reconfiguration rested in the example it provided for other components of the railroad system.[25]

System-Wide Innovation: Imperatives of Politics

With publication of Braman Adams' book *The Block System* in 1901, knowledge of the Pennsylvania Railroad's automatic block signaling system became readily available. Adams devoted much of his study to the recent developments at the Pennsylvania and left no doubt that they represented the finest system in the land. At virtually the same moment, block signaling became a subject of political reform as the mounting pressure for federal regulation of railroad rates swept along with it a movement for government policies to prevent accidents.

This combination of events marked the beginning of a second effort of systems' reconfiguration associated with block signaling. For the next dozen years, through political action the American people would seek to revise the system for governing the movement of passenger trains in the United States.

The block system provided the lever for that effort. It opened an opportunity for regulation and established a basis for discussing train-handling and government's proper role. But the block system did not yield a simple, straightforward course for this system-wide reconfiguration to follow because the block system meant many different things to many different people. In the eyes of many Americans, it referred to the practices at the Pennsylvania that appeared to provide absolute safety through technological means. For others, it referred to the space-interval method of operations, which could be accomplished very simply. For some railroads, including the Pennsylvania, the block system was an economical tool even when it required complex equipment. For others it was a costly extravagance in any form. And in the opinion of some influential government officials, the block system in any manifestation was merely a first step toward attaining safer governance of trains. They argued for a system of regulation that would provide the federal government with the means to supervise virtually every aspect of railroading that affected the safe movement of trains. This section of the chapter will trace their efforts.

Law Introduced for Block System

Sentiment for regulation of railroad train-handling built rapidly during the first decade of the twentieth century. In 1901, Congress passed a law requiring railroads to submit statistics regarding accidents. This measure coincided with the date of compliance for the Safety Appliance Act, a law passed in 1893 that mandated the use of self-couplers and automatic brakes on all trains. The ICC published its first accident statistics in its Report of 1902, calling them "the first authentic record [in the matter of collisions and derailments] relating to the railroads of the whole country," and observed that "the publication of official data has drawn renewed attention to the subject." The commissioners said the block system provided the best remedy for collisions.[26] The following year Representative John J. Esch of Wisconsin introduced legislation to require the block system. The ICC immediately supported this proposal, which basically copied the British law of 1890, and added a recommendation that it be given power to supervise the railroads and insure compliance. President Theodore Roosevelt called for passage of the legislation in his annual messages to Congress in 1904 and 1905.[27] In its reports for those years the ICC lobbied emphatically for the bill and also asked for the power to investigate accidents.[28]

These efforts to secure regulation of safety coincided with an extended debate over the role of the ICC in setting railroad rates. During the late nineteenth century, the Supreme Court had essentially deprived the commission of authority in this vital matter. The judicial challenge to regulatory authority sparked enormous controversy and fomented a renewed cry for legislative action. In 1901, Congress tabled a bill that would have provided the ICC with limited rate-making powers. However, two years later it gave the commission authority to prosecute railroads that offered rebates or other discounts from their published fares. Congress then proceeded to consider legislation that would permit the ICC to set maximum rates and grant it more authority to investigate and monitor railroad finances. When the House and Senate split on this matter, the President decided to strongly support a measure giving the ICC broad authority. At the same time, he endorsed the regulation requiring block signals.[29]

In 1906 Congress took action on both the rate question and the issue of block signals. With regard to rates, it passed the Hepburn Act that basically advocated Roosevelt's positions on commission oversight of rates and finance but left open the possibility that courts would continue to review ICC decisions. On the matter of signals, Congress resolved that the ICC should investigate block signals and "appliances for the automatic control of trains" and make recommendations regarding legislation.[30] These actions did not put the matters of rate regulation and safety to rest. Rather, they established an institutional framework for further debate and action. In each case, Congress had bolstered the administrative responsibilities of the ICC but had not precisely defined its authority.

The political events of 1906 marked a watershed in the development of railroad signaling and methods of handling trains. By dealing with signaling in a similar fashion to the question of rates, Congress almost certainly had drawn greater attention to signals than they would otherwise have received. Moreover, Congressional actions insured that the ICC would now play a more prominent role in deciding how best to safely move trains. Government had entered the American train-handling system. Precisely how radical a reconfiguration that would entail remained to be seen.

Conflicting Approaches to Safety

That question's answer would be largely determined by the latitude and authority Congress ultimately allowed the ICC. When it received the resolution of 1906, the commission had for many years been trying to move beyond an approach to safety based on laws requiring particular technological devices and substitute for it one grounded upon supervision and inspection. Commissioners believed the experience gained in implementing the safety appliance law had conclusively demonstrated the wisdom of their efforts. Though this act had required self-couplers and brakes by 1902, the number of deaths and injuries due to coupling and braking had not diminished until after 1904 when government inspectors had begun visiting railroads and supervising maintenance and use of the required equipment.[31] As often as possible, the commission tried to use the limited powers of inspection it had obtained from the safety appliance law. It regularly attached summary statements from its air-brake and coupler inspectors to its annual reports and drew attention to inspectors' observations about safety issues. Often these sections of the reports highlighted the importance of personnel, training and discipline in the safe conduct of railroad transportation. The commission continued to seek the right to investigate accidents and occasionally reported informally on wrecks that had attracted widespread public attention.

The resolution of 1906 held ambiguous implications for this approach to safety. Congress had given the commission authority to investigate, report and recommend on a subject it had previously addressed only informally. That subject, moreover, touched upon the very matters of rules, discipline and operational procedures that commissioners had come to see as essential to safe transportation. The ICC might conceivably use the resolution to obtain additional supervisory authority over railroad operations. But the resolution from Congress also contained strong overtones of the convential, technology-centered approach to safety regulation. It referred to "block signals and appliances for the automatic control of trains." If construed narrowly, the resolution might divert the ICC away from its emerging emphasis on inspection and supervision.

The conflict between these approaches to safety was apparent from the moment the ICC first responded to the block signal resolution in 1907. It continued to influence all actions and discussions on government regulation of train-handling for at least another six years. Nothing demonstrated this more clearly than the experiences of the commission when it tried to investigate "appliances for the automatic control of trains." Congress apparently had in mind devices that would stop a train automatically if the engineer passed a red signal. From the perspective of those who viewed safety as the substitution of mechanical devices for human judgment, automatic stops were the next logical extension of the automatic block system. "Every great accident due to an engineman's fault," skeptical commissioners wrote in one report, "is followed by a strong public demand for the introduction of automatic train stopping devices."[32] Inventors responded with countless arrangements they claimed would do the trick. Though a few urban transit systems had tried such devices on an experimental basis, none had yet proved satisfactory. Too often these tempermental appliances had stopped trains inadvertently, creating more danger than they had been designed to prevent. Railroads such as the Pennsylvania placed far more hope on "surprise tests" than on automatic stops and the ICC concurred.[33]

Faced with evaluating numerous devices whose merits had not yet been demonstrated, the ICC requested special funds to conduct tests and in July of 1907 it received authorization from Congress to create the Block Signal and Train Control Board to carry out the task. This action opened the floodgates. Inventors not only submitted innumerable designs for automatic stops, they also persuaded Congress to require the Signal Board to examine *any* prospective invention on railroad safety. The board received hundreds of proposals every year and responded to them all, even though in a typical year it found fewer than a dozen to possess "sufficient merit to warrant further test."[34]

Persistence of Technological Remedies

While the ICC labored under this unsolicited burden, it struggled to maintain public focus on the block system and the broader issue of safety. Even when the commission discussed the block system, however, it often had difficulty getting past the matter of technology. Many people viewed the block system not as a method but as a set of novel devices. Even the resolution from Congress had referred to "block signals," not the "block system." The commission pointed out this discrepancy to Congress in its initial response and in subsequent reports always spoke of the block system. Commissioners directed attention to the role of management and workers in administering the block system and stressed the importance of proper training and supervision of signalmen.[35]

Despite its intentions, the ICC in many ways reinforced the tendency to think first about technology. In response to the resolution asking for an investigation of block signals, the commission in 1907 had two consultants produce a report on the subject. One was Adams and the other was C. C. Anthony, who for the previous eleven years had served as supervisor of signals at the Pennsylvania. These authors carefully defined the block system as a method and described its various forms and the extent of its use in actual practice, but perhaps inevitably, they dwelled at length on the most sophisticated installations. The influence of developments at the Pennsylvania loomed especially large. Because the report was an official government document that could not advocate particular devices, the authors referred to these developments in generic terms and used the passive voice when discussing them, which heightened their significance. Rather than portraying automatic block signals as a specific response to a particular set of circumstances, the report characterized them as "the highest exemplification of the art" and noted that they "are superseding the telegraph block system to some extent, and seem destined to do so increasingly in the future."[36]

To make the strongest case possible for a mandatory requirement while retaining their tone of impartial investigation, the authors tried to identify as many arguments in favor of adoption of the block system as they could. This led them into detailed discussions of costs of various systems and extended ruminations about the potential of block methods to increase capacity and provide other economic benefits. Consequently, the report ended up possessing an encyclopedic quality. It lacked the morally compelling sense of urgency a reformer might have brought to the safety issue. Its thorough, considered judgements of the economic costs and benefits of various installations could hardly have diverted attention away from technology nor built support for the ICC's preferred approach to regulation.

Perhaps the strongest points made in favor of inspection and supervision in the consultants' study and subsequent ICC reports were those regarding personnel. Often these arguments came in the form of a comparison with Great Britain. By all accounts, British signalmen were better trained and more experienced than their American counterparts and they came under supervision of the British Board of Trade.[37] But here the ICC encountered a dilemma. In emphasizing the need for a well-trained, disciplined workforce subject to government oversight, the ICC also lent support to those who saw automatic technology as the key to improving safety. In appealing for an approach more like that of the British, the ICC was challenging the prevailing faith in technological remedies. "In this country we spend millions in an endeavor to make our apparatus fool proof," a prominent American signal engineer said of the American attitude," while in England they spend hundreds to eliminate the fool, and appear to get better results."[38]

This dilemma on labor was brought home to the ICC when Congress passed the Hours of Service Act in 1907. This law stipulated that signalmen could not work longer than nine hours at a stretch unless the railroad received special exemption from the ICC. At the time railroads routinely manned stations in two twelve-hour shifts, so it came as no surprise when over fifty companies requested exemptions when the law took effect in March 1908. The ICC, which had campaigned for the law, summarily denied them all. Within months, the Signal Board detected a dramatic drop in the rate of adoption of the block system. The number of miles operated under the manual block system actually decreased for the year as some roads switched to automatic methods to save labor costs. An attempt to act on labor's behalf had worked against securing use of the block system and had shifted attention toward automatic technology.[39]

For those who had hoped the resolution on block signals would increase government authority over the movement of trains, the experiences of 1906-1909 were a great disappointment. They had fallen victim to the American predilection to focus on technological devices rather than on the entire system in which those devices functioned. Ironically, the developments at the Pennsylvania, which safety advocates could only have praised, had worked against their efforts to achieve more systematic reform.

Ambiguous Outcome of Reform

These frustrations spilled over in the Signal Board's report for 1911 when three of the four members argued that "the time has come in this country to inaugurate a system of supervision over interstate roads somewhat similar in character to that now administrated through the British Board of Trade." These board members asked for laws that would give a supervisory agency regulatory authority over "the details of construction, maintenance and operation of all interstate roads, so far as concerns safety of railroad travel and employment." Clearly frustrated with its experience of the previous "four and one-half years of special study, investigation and observation," the board stressed that the actual study of devices and methods should be left to railroads. The ICC should watch over actual practice. "The railroads are well organized to represent their own interests, as are also their employees," the report stated in its ultimate justification for its recommendation, "but the public must look to a governmental agency to protect its interests."[40]

The lone dissenter from this position was Braman B. Adams, the trade journal editor, who had joined the Signal Board after his initial stint as a consultant. Adams submitted a minority report in which he criticized the idea of an inspection board and defended an approach to safety that was based on the power of publicity to hold managers accountable. "Beyond publicity we enter an untried field," Adams warned. "If and whenever an agent of the government decides what a railway officer shall do, the State assumes some degree of responsibility for the acts of such officer." Adams argued that any experts employed by the government in such a broad supervisory capacity would need a level of experience commensurate with that of the general manager of a major railroad. Until the ICC hired such men and trained them, a prospect Adams clearly considered highly unlikely, Congress should simply require the block system.[41]

By the time this difference of opinion was expressed, a decade of debate over accident prevention and the block system had revealed several policy alternatives. One approach advocated by a significant segment of the public was to require that the railroads adopt particular technologies. Most of those responsible for administering regulation no longer viewed this approach as sufficient to insure safety. In the case of the block system, they saw the focus on technology as distracting from real reform. These regulators preferred a law that focused instead on the block method. Above all they sought broad supervisory powers and administrative authority. Adams, who from the perspective of his fellow board members had more sympathies for the railroads than for the public, opposed both regulation mandating specific technologies and expansion of government supervisory authority. He sought to restrict the role of government to that of publicist. In supporting a law requiring the block method, Adams moved only slightly away from that position. As he well understood, the term block system covered a range of options including permissive application. Railroads would retain broad discretion over their operations.

In its actual policies, Congress had pursued an approach of intensified publicity. It had allowed the ICC to gather information about accidents, survey existing practice on the block system, and evaluate and publicize technical developments. Congress had not passed legislation requiring the block system. Nor had it significantly expanded the power of the ICC to inspect and supervise ordinary practice.[42]

The arguments submitted by the majority of the Signal Board did not prompt Congress to alter its course significantly. Congress closed down the Signal Board after 1912 and created a Division of Safety, which took over responsibility for tasks already authorized under existing statutes, such as gathering statistics, inspecting devices required under the safety appliance law and investigating accidents. Congress dropped the routine evaluation of prospective safety devices but in 1913 authorized an emergency allocation to renew studies of automatic stops. Despite a fervent plea from the Signal Board, which claimed in its final report that "railways have been slow in adopting the block system" and labeled the absence of a requirement a "failure of government," Congress did not require the block system and continued instead to rely on its expanded powers of publicity.[43]

Summary and Conclusions

Between 1901 and 1915 far more railroads came to use the block system in their passenger operations than before. In 1901 Adams reported that less than 27,000 miles of road were being operated under the block system and of those fewer than 2,300 miles had automatic signals. Many companies used the method permissively and some lines employed the block system only during foggy weather or other exceptional circumstances.[44] By 1915 those figures had risen to 96,609 miles under the block system and 29,864 miles with automatic signals. "The block system is in use on the entire passenger mileage of the important roads of the country," declared the ICC that year, "and the activities of these roads in signal work now consist principally of substituting modern for antiquated equipment and improving operating conditions and practices."[45]

Conversion to the block system had not, however, entailed a wholesale switch from one system to another. In the end, fifteen years of experimentation by individual lines and another fifteen of political haggling had resulted in a hybrid system for governing the movement of trains. Over sixty percent of the country's railroad mileage still operated under the older dispatching system and those that used the block system employed it in a wide variety of forms. Few railroads had reconfigured their systems as radically as the Pennsylvania. Government had increased its influence but its involvement did not come in the form of blanket requirements or sustained supervision. Rather, it left railroads free to operate as they pleased, now knowing that government investigators might examine and publicize their mistakes and that ignorance of the block system was no excuse.

In terms of the nomenclature introduced at the outset of this paper, the long course of innovation in railroad signaling had culminated in a less thorough systemic change than first was apparent. Signaling remained an embedded system of substantial indeterminancy, which in turn helped sustain a high level of differentiation in the railroad system as a whole. The regulatory system had expanded and secured a more prominent supervisory function, but its circumscribed role kept it from implementing more comprehensive changes that might have reduced the level of indeterminancy throughout the hierarchy. Change had occurred largely within the constraints of the established systems and the change's net effect was to preserve sufficient indeterminancy so that subsequent innovation would also likely fall within the accepted range of tolerance in those systems.

That these systems ultimately resisted more radical change should not surprise us. It is almost axiomatic of systems that they should do so. In the case of a system embedded within many others, we might well anticipate still greater built-in resistance to change. Yet interestingly, the case of railroad signaling offers scant support for such an assumption. Reformers narrowly missed instituting a far-reaching reform in the way the movements of all trains in the United States were governed. Had the bold innovations of the Pennsylvania Railroad not been embedded in the larger railroad system, this could hardly have come so close to happening. Those innovations came in response to a particular set of circumstances — above all the congestion of the Pennsylvania's lucrative lines. Few other railroads would have found reason to pursue them on their own accord. Because those innovations were embedded in larger systems, regulators and safety advocates came close to making them universal.

Changing an embedded system can be explosive. It can set off a cascade of reactions throughout the hierarchy. One can imagine a change so powerful that it compels all of the higher-order systems to accommodate it. (In railroading dieselization comes to mind.) In the case of signaling, the explosive qualities were more subtle. Their source lay in the high degree of indeterminancy and differentiation which gave rise to many problems and many agendas. Those agendas intersected but were not identical. At some points they reinforced one another, at others they worked in opposition. The means of solving one problem (congestion) became the ends for those looking to solve another (safety). A broadly felt desire to improve safety became something of a pawn in a struggle to establish fundamental regulatory principles. This confusing climate held the potential for change of an order that those working on the embedded system could hardly have imagined.

Notes

1. "19th Annual Report of the Interstate Commerce Commission" (hereafter, ICC Report), 12/14/05, p. 77.

2. An influential ICC investigation, for example, attributed the change simply to "mixed motives." "Report of the Interstate Commerce Commission on Block Signal Systems and Appliances for the Automatic Control of Railway Trains," 2/23/07, U. S. Senate, 59th Congress, 2d Session, Doc. No. 342 (hereafter, ICC Block Signal Report), p. 22.

3. Braman B. Adams, *The Block System of Signaling on American Railroads* (New York, 1901), pp. 7-23. In 1901, the 105-mile stretch from Philadelphia to Harrisburg contained 51 sections.

4. ICC Block Signal Report, pp. 6-7.

5. Report of the Committee on Signals, 1/14/80, Pennsylvania Railroad Records (hereafter, PRR Papers), Association of Transportation Officers Files (hereafter, ATO), Hagley Museum and Library. On interlockings in general, see James Brandt Latimer, *Railway Signaling in Theory and Practice* (Chicago, 1909), pp. 35-247.

6. "Report of the Committee on Interlocking and Block Signals," 10/24/1893, PRR Papers, ATO Files, Box 1, File 16.

7. This phrase comes from a 1905 letter from General Manager W. W. Atterbury to President A. J. Cassatt explaining the function of automatic signals. "As the function of the automatic signal is to pass the maximum traffic over the road at the greatest speed consistent with safety," Atterbury wrote, "it follows that each train should be kept at its maximum speed over as great a portion of the road as possible." W. W. Atterbury to A. J. Cassatt, 4/4/05, PRR Papers, ATO Files, Box 7, File 413. The report of 1893 stressed that the general plan for signaling applied to "lines where a large traffic is handled and where speed and safety are the ends sought."

8. "Report of the Committee on Interlocking and Block Signals," 11/21/1894, PRR Papers, ATO Files, Box 2, File 35.

9. The view of distant signals as tools to avoid slacking speed differed, of course, from that often held by safety advocates, who viewed them as an added layer of protection. When President A. J. Cassatt suggested in the wake of an accident in 1905 that engineers be required to slow down at caution sig-nals, General Manager W. W. Atterbury resisted. He pointed out that the purpose of signals was to keep speeds high. See note 7 above.

10. "Report of the Committee on Interlocking and Block Signals," 10/24/1893, PRR Papers, ATO Files, Box 1, File 16.

11. N. Trump (Chairman, Signal Committee) to S. M. Prevost (President ATO), 10/29/1896, PRR Papers, ATO Files, Box 1, File 16.

12. "Report of the Committee on Interlocking and Block Signals," 10/24/1893, PRR Papers, ATO Files, Box 1, File 16.

13. "Report of the Committee on Interlocking and Block Signals," 11/21/1894, PRR Papers, ATO Files, Box 2, File 35.

14. N. Trump (Chairman, Signal Committee) to S. M. Prevost (President ATO), 10/29/1896, PRR Papers, ATO Files, Box 1, File 16.

15. "Report of the Committee on Train Rules," 10/29/1898, PRR Papers, ATO Files, Box 1, File 6.

16. Adams, p. 165.

17. Steven W. Usselman, "From Novelty to Order: George Westinghouse and the Business of Innovation during the Age of Edison," *Business History Review* 66 (Summer 1992): 251-304.

18. Latimer, pp. 345-355.

19. "Report of the Special Committee on Signal Practice," December 1905, PRR Papers, ATO Files, Box 6. Minutes of the ATO, 11/26/07, p. 4, PRR Papers, ATO Files.

20. "Report of the Committee on Interlocking and Block Signals," 11/21/1894, PRR Papers, ATO Files, Box 2, File 35.

21. The books were those of Adams and Latimer, respectively. A number of other textbooks appeared over the next dozen years.

22. Ironically, within a few years telegraph operators at the Pennsylvania would fall victim to a systematic effort by management to replace them with telephones, a technical innovation that had not been developed expressly to displace them but which was adopted to save labor costs. "Report of the Committee on the Telegraph," 11/22/1897, PRR Papers, ATO Files, Box 5. "Joint Report of the Committee on Train Rules and the Committee on the Telegraph," 10/30/05, PRR Papers, ATO Files, Box 6, File 325. G. W. Creighton (General Superintendent of the Eastern Pennsylvania Division) to W. W. At-terbury, 6/4/08, PRR Papers, ATO Files, Box 5, File 263. J. B. Fisher (Superintendent of the Telegraph) to W. W. Atterbury, 5/26/08, PRR Papers, ATO Files, Box 5, File 263.

23. Adams, pp. 23-24. ICC Block Signal Report, p. 10.

24. G. W. Creighton to W. W. Atterbury, 1/28/08, PRR Papers, ATO Files, Box 9, File 493.

25. Some of those other components could be found at the Pennsylvania itself, since the automatic block system was still not used on some divisions in 1908. Even on divisions with the automatic block system, moreover, freight trains were still permitted to enter an occupied section after stopping if the preceeding train had also been a freight train.

26. 16th ICC *Report*, 12/15/02, pp.64-69.

27. "Block Signals and Appliances for the Automatic Control of Railway Trains," House Report No. 4637, 59th Congress, 1st Session. ICC Block Signal Report, p. 25.

28. 17th ICC Report, 12/15/03, pp. 102-4. 18th ICC Report, 12/19/04, pp 98-99. 19th ICC Report, 12/14/05, pp.77-78.

29. Stephen Skowronek, *Building a New American State: The Expansion of National Administrative Capacities, 1877-1920* (Cambridge, 1982), pp. 248-256.

30. ICC Block Signal Report, p. 25. Skowronek, p. 256.

31. 19th ICC Report, 12/14/05, pp. 168-173.

32. ICC Block Signal Report, p. 26.

33. For attitudes at the Pennsylvania, see G. W. Creighton to W. W. Atterbury, 1/28/08; R. M. Patterson (Superintendent of Freight Transportation) to Atterbury, 2/18/08; and Thomas E. Reilly (Superintendent of the Telegraph) to Patterson, 2/10/08; PRR Papers; ATO Files; Box 9; File 493. For ICC praise of surprise tests as "progressive," see 16th ICC Report, 2/15/02, pp. 303-311 and ICC Block Signal Report, p. 17. On automatic stops, see "Reports of the Block Signal and Train Control Board to the Interstate Commerce Commission" (hereafter, BSTC Board Reports), esp. the 4th report, 12/26/11.

34. BSTC Reports, esp. the 4th report, 12/26/11.

35. ICC Block Signal Report, esp. pp. 14, 17, and 28. 23d ICC Report, 12/21/09, p. 50.

36. ICC Block Signal Report, p. 22.

37. "The facts of the accident records justify a strong presumption that American signalmen are not so carefully selected nor so well trained as those of England. The average signalman in America is young, and has had, proba- bly from six months to two years instruction — not systematic instruction — under another signalman, whose superiority to the student is entirely due to what he has learned by experience and not at all to methodical and authorita- tive instruction. The average block signalman in England, on the contrary, has served as such from five to twenty-five years and has been through a long course in a signal cabin as a "booking boy," or as an assistant, before being trusted with full charge of the block signals. This difference in personnel of the signalmen of the two countries undoubtedly explains in large measure the nearer approach to perfection of the block signal service in England." ICC Block Signal Report, p. 14.

38. Latimer, p. 350.

39. 21st ICC Report, 12/23/07, p. 132. 22d ICC Report, 12/24/08, p. 307. 23d ICC Report, 12/21/09, p. 49. Latimer, p. 348.

40. 4th BSTC Board Report, 12/26/11, pp. 12-18. Quotes from pp. 14 and 16.

41. Ibid., pp. 18-22. Quote from p. 20.

42. In 1910, Congress granted the ICC authority to investigate accidents in addition to collecting accident statistics. In subsequent years, however, commissioners complained that the available funds permitted them to carry out only a few accident investigations each year. See "Reports of the Chief of the Division of Safety to the Interstate Commerce Commission" (hereafter ICC Safety Reports), 1913-1915.

43. ICC Safety Reports. Final BSTC Board Report, 6/29/12, p. 22.

44. Adams, pp. 170-171.

45. ICC Safety Report (1915), pp. 18-19.

PART TWO

Crossing Territorial Borders

6

Transformation Through Integration: The Unification of German Telecommunications

Tobias Robischon

The process of German unification resulted in radical changes to both German telephone systems. Not only did it alter their political and geographical environment, but it also defined new requirements for their performance. When the borders opened up, the demand for telecommunications between East and West skyrocketed. The outdated and undersized East German telephone system had to cope with an enormous demand for extensions as well as modern telecommunication services. The changes in their environment and the new demand forced actors in East and West to decide on a possible transformation of the system. This decision-making process is the subject of this chapter.

Telecommunications in the German Democratic Republic

The telecommunications system of the German Democratic Republic (GDR) was organized in the traditional form of a post, telegraphy and telephone administration. Legally based on a wide-ranging monopoly, a governmental agency called Deutsche Post operated the public telecommunications network as well as the postal services. Deutsche Post in turn was headed and managed by a Ministry of Postal Services and Telecommunications.

Although the Deutsche Post had its own building and construction units and production of radio equipment on a small scale, the telecommunications system of the GDR was not fully vertically integrated.[1]

The GDR's telecommunications industry consisted essentially of one conglomerate, the Kombinat Nachrichtenelektronik, with eighteen production sites scattered over the country and 36,000 employees.

The budget of the Deutsche Post was part of the GDR's national budget. Therefore, the Deutsche Post depended fully on the funds allocated to it by the GDR government, which actually was the central committee of the ruling socialist party SED (Sozialistische Einheitspartei Deutschlands). Surplus made by the Deutsche Post went almost entirely to the general budget, while its own budget consistently remained small.[2] Since it was not possible to build up savings for future investments, the Deutsche Post had no leeway for any self-regulated investment activities.

In addition to its small share of the national budget, its low priority in the distribution of goods by the material economic balancing system restricted the Deutsche Post's activities. The material balancing was done by the central economic planning commission, a group under the direct control of the SED's Central Committee.[3] The investment part of Deutsche Post's budget was too limited to cover the cost of both the network's expansion and the necessary repairs and replacements. Since the need for additional lines was enormous, priority was given to their creation. This was mostly accomplished by the division of existing main lines into party lines. Because the budget did not permit the replacement of outdated equipment, the share of worn-out material steadily increased.

Because the investment behavior of the Deutsche Post was closely coupled with the SED's policy goals, the development of the telecommunications infrastructure was also coupled to their policy. Telecommunications scored very low, however, on the SED's list of political priorities. Despite long waiting lists, the official doctrine considered private telecommunications unnecessary and not essential to an adequate standard of living. While this policy may be mostly due to the GDR's lack of funds, the SED's desire to limit and hinder contacts to the west and its aim to control all citizens' social activities surely played a part.

The Deutsche Post and the Ministry of Postal Services and Telecommunications, believing public telecommunications should receive more attention and resources, considered the SED's policy to be systematic discrimination. Nevertheless, the planning division of the ministry developed numerous ambitious projections, e.g. to nearly double the expansion rate of the public network in the 1990 to 1995 planning period. Their plans also included the idea of importing a data transmission network and beginning to digitalize the existing telecommunications infrastructure. Even if such projections had been realized, the West

German 1990 supply level would not have been reached before the year 2060.

Although the total output of the Kombinat Nachrichtenelektronik (i.e. the combined telecommunications industry of the GDR) was a sizable half million network access units a year, the Deutsche Post could take almost no advantage of these capacities. The largest part of the production had to be exported, mostly to the USSR. The Deutsche Post received only seven percent of the switching equipment, while other network operators, such as the GDR government, the secret service and the military received eleven percent.

Nearly all telecommunications equipment manufactured by the Kombinat Nachrichtenelektronik was based on analog technologies. Only a small-scale production of low-capacity digital systems was possible. Generally speaking, digital telecommunications equipment was unavailable to the Deutsche Post — and could not be purchased within the COMECON ("Council for Mutual Economic Aid" of the Eastern bloc) or imported from western countries. The GDR's shortage of hard currency as well as COCOM regulations (the western boycott list of high technology products) restricted its ability to import.

The telecommunications infrastructure of the GDR was determined by this situation. Its telephone distribution rate of eleven main lines per 100 inhabitants was low compared to other industrialized nations, where the rate is forty to sixty.[4] The public network was outdated because many exchanges dated back to the 1920s and the latest technology had been introduced in the early seventies. The small public telephone network had only 1.8 million subscribers connected, some sixty percent of them by party lines. Due to the lack of resources, as many extensions as possible had been connected to the network. Large private branch exchanges had to make up for the limited number of network accesses.

Due to the bad condition of the public network, separate networks of higher quality had been set up to fulfill the needs of state agencies and industry. These networks had grown to remarkable proportions, comprising several hundred thousand connections.

Since network growth could never meet the growth in demand, more than 1.2 million applications had piled up in the Deutsche Post offices by 1989. The average waiting time for a phone was ten to twenty years, which even exceeded the waiting time for a car. The high, unmet demand was nearly entirely private, while the requirements of state and industry had been covered. Here, the results of the SED policy not to meet the private demand for telecommunications become very evident.

Since relevant political and economic actors of the GDR such as the socialist party, the secret service, the state administration and important branches of the industry operated their own telecommunication networks, the political position of the Deutsche Post was weak. Too few politically influential actors depended on improvements in the public telecommunications network. The Central Committee's policy not to expand the public network for private demand thwarted the Deutsche Post projections. But even without this policy, the GDR's lack of economic resources, its lack of modern technology, its shortage of hard currency and its many obligations to deliver to the USSR would have made improvements in the GDR's telecommunications system more than difficult.

Telecommunications in the Federal Republic of Germany

Only a few months before the opening up of the borders in November 1989, the telecommunications system of the Federal Republic of Germany (FRG) had undergone a thorough reform.[5] Although the motto of the reform was to introduce competition as a rule and to retain the monopoly as an exception, in practice competition still is the exception. The basic transmission infrastructure and the telephone service, which account for seventy-six percent of the network operators' turnover, are still under the state monopoly.[6] The idea behind keeping a monopoly in telephony was to use the returns of monopoly services for subsidies to the deficit postal services and expenses due to the monopoly's infrastructural obligations. Since the reform act was not precise in drawing the borders of those monopoly rights and defining which services the monopoly operator was obliged to provide, further liberalization and disentanglement of the telephone monopoly were expected.

The West German Post and Telecommunications organization (the Deutsche Bundespost), one of the country's biggest corporations, had been split up by the 1989 reform act into three separate companies for postal services, banking and telecommunications.[7] Its telecommunications branch alone, the Deutsche Bundespost Telekom (DBP Telekom) had 212,000 employees in 1990. It accounted for a turnover of about DM 40.6 billion and reached a gross surplus of DM 7.4 billion. The company spent DM 19.2 billion on investments in its network.[8]

The task of the Federal Republic's Ministry of Post and Telecommunications is to open and regulate the telecommunications markets and control the market behavior of the Telekom monopoly. Since the ministry is at the same time the owner of and superior authority for the Deutsche Bundespost, it is able to direct the policy of DBP Telekom.

An important group of actors in the West German telecommunication system are the telecommunication equipment manufacturers. Because of their close relationship to DBP Telekom, the largest firms have received the nickname "court suppliers."[9]

For decades, the telecommunications industry and the Deutsche Bundespost cooperated closely in developing the West German telecommunications infrastructure. The practically competition-free national market has not only ensured a high sales potential, but it has also been of strategic relevance to the industry. Although the market is too small to completely cover the high development costs of digital systems, it guarantees the necessary sales base for their development. Since, mostly for technological reasons, long lasting supply relationships are typical for the telecommunications sector, the 1989 reform act was not expected to change this situation.

Users are not well represented in the system because of two reasons. First, there are no organizations advocating the interests of residential users. Second, the associations of trade and industry representing the interests of commercial users represent the interests of the telecommunications manufacturers as well.

The political parties can exercise an influence on DBP Telekom policies by means of the National Parliament's Committee on Postal Affairs or the Infrastructural Council. The latter is an advisory body composed of political appointees from the federal states and the national parliament. But the political opinions expressed in these committees are quite diverse and range from calling for the return to the old monopolistic system to demanding a quick dissolution of all monopoly rights.[10]

The old Federal Republic was not among the top countries regarding the provision of telecommunications, but it was nearly completely supplied with telephone service. The telephone distribution rate was thirty-nine main lines per 100 inhabitants. Connecting 30.6 million main stations, the West German network was more than sixteen times the size of the East German network. A systematic digitalization of the network had been started in the mid-eighties. By 1990, about eighty percent of the long-distance exchanges were digital. On this basis, subscribers could be connected to ISDN (Integrated Services Digital Network),[11] which had been introduced to the network in 1988.

Considering the nearly full provision of the country with telephones, the Deutsche Bundespost tried throughout the eighties to introduce new services such as videotex and ISDN to open up new opportunities for growth. The long-term target of the Deutsche Bundespost is a universal high-capacity network which could integrate all media and telecommunication services, such as the transmission of voice and picture telephony, the distribution of television and the exchange of data.

November 1989 to March 1990:
Inter-German Cooperation and Technical Compatibility

The opening up of the Berlin wall and the downfall of socialism in the German Democratic Republic changed the situation on both sides of the wall.

The Deutsche Post was primarily occupied with keeping up its everyday work in a revolutionary situation. At the same time it was confronted with protesting East German citizens who demanded better service and phones that they had been waiting for for years. As already mentioned, the most acute shortage in East German telecommunications was in new extensions for private households. Since citizens' wishes now mattered politically, the East Berlin telecommunications ministry aimed at a rapid increase in the number of telephone subscribers. There were very few lines to West Germany but this was not regarded to be a major problem.[12]

The GDR's telecommunications ministry tried to improve the situation by utilizing the Kombinat Nachrichtenelektronik's production capacities. But this did not prove to be feasible. Although the Kombinat Nachrichtenelektronik was subordinated to the telecommunications ministry, the production of the Kombinat could not be mobilized for an expansion of the public network. The commando system of the GDR economy had ceased to work. The switch to market economics had led to a disintegration of the Kombinat, whose parts now strived for economic autonomy.

The western DBP Telekom had to cope with a skyrocketing demand for telecommunications to East Germany.[13] Since there were too few lines to East Germany to satisfy demand, the many attempted calls occasionally blocked parts of DBP Telekom's network. Because the many attempted calls used the network extensively without generating fees, DBP Telekom was interested in an increase in lines to East Germany. This would have helped but only to a certain degree. The eastern network was too small to handle much incoming traffic, so DBP Telekom could not take advantage of a large increase in trunk lines to East Germany. This increase in trunk lines without a proportional expansion of the eastern long-distance network would only jam the East German instead of the West German exchanges.

As a result of the pressing necessity to coordinate the inter-German telecommunications traffic after the opening of the border, a number of cooperation initiatives were launched between the East and West German ministries for telecommunications. Interministry working groups agreed on additional trunk lines and other improvements in East-West traffic. At the end of January 1990, the working groups were institutionalized with the foundation of a "Joint Government Commission." It was understood by the Federal Republic's Ministry of Telecommunications that it was necessary to improve the Deutsche Post network if lasting improvements in East-West telecommunications were to be achieved.

Following the usual inter-German policy pattern of facilitating interpersonal communications and contact, the Federal Republic's telecommunications ministry supported the modernization of the eastern network as a necessary technical prerequisite. To achieve lasting improvements in East-West traffic, the Federal Republic's telecommunications ministry employed the standard means of the Postpauschale.[14] A DM 50 million increase, raising the payment to a sum of DM 250 million, was combined with the stipulation to use the money only for improving telecommunications and postal services. In previous years, the Postpauschale had been used by the GDR to support its national budget instead of its postal services or its telecommunications network.

In addition, the western ministry promoted the Federal Republic's telecommunications industry within cooperation talks and maintained that the best way to improve East-West telecommunications would be to reach compatibility of the two networks. This would necessitate introducing West German telecommunications' technology in the GDR.

Upon receiving the DM 250 million Postpauschale, the Deutsche Post had for the first time much hard currency at its disposal. This windfall opened up completely new possibilities. Long-planned goals could now be realized with old projections pulled out of the drawer. Consequently, the first projections of the Deutsche Post aimed at a larger network for the GDR and a steep increase in subscribers. The Deutsche Post did not plan to build a close connection to West Germany, nor did it intend to change the network's existing structure. Following its old projections, the Deutsche Post planned a proportionate expansion of the network at all hierarchical levels and 100,000 additional main lines in 1990.

But within the "Joint Government Commission" talks, Deutsche Bundespost engineers gave different advice to the Deutsche Post planners. The Deutsche Post and its ministry, who were determined to bring their network up to date, quickly accepted the counseling and readily approved the proposals. This was not very hard for them, as Deutsche Bundespost technologies had always been their technological model.[15]

The commission's working groups, consisting of experts from the Deutsche Post and the Deutsche Bundespost, conceived how to fulfill both goals of more GDR main lines and a closer connection to West Germany. A digital overlay network was to guarantee the interconnection of the two networks from the start. The network would enlarge the Deutsche Post's long-distance network and link it closely to the West. It would run in a big loop through the northern and southern GDR to the Federal Republic. Since the loop touched some of the main nodes of the Deutsche Post network, traffic could flow directly to the FRG and avoid the bottleneck of international exchanges. In addition, towns along the loop could be provided with a limited number of new connections. By 1991, about 100,000 new subscribers, primarily businesses, could be connected. For both network operators, the overlay network solved their problems. While the Deutsche Post could partly meet the demand for additional phones in East Germany, both operators could benefit if the bottleneck in East-West traffic was overcome.

The first cooperative activities between the East and West German telecommunication industries were started in early 1990. By then, the Kombinat Nachrichtenelektronik had actually begun to fall apart. Each plant tried to ensure its survival on its own by cooperating with western companies. At the same time, West German corporations such as Siemens or SEL (a telecommunications manufacturer) tried to come in contact with East German companies. They hoped to gain access to the East European markets, which promised to be very profitable, by investing in East German companies. The West German telecommunications corporations acted quickly in an effort to reach favorable strategic positions on these markets. In March 1990, SEL was the first corporation to found a joint venture with a former Kombinat Nachrichtenelektronik company. Shortly afterwards, it was followed by Siemens, Bosch and others.

At the Leipzig spring fair in mid-March 1990, the GDR's Ministry of Postal Services and Telecommunications opted for the proposed modernization through an overlay network it commissioned the West German corporations Siemens, SEL, PKI and ANT to construct. The cooperative activities and the willingness of the corporations receiving the orders to invest in the GDR's telecommunication industry accelerated the decision. In addition, the two German telecommunication ministers signed an agreement that they aimed for full compatibility of all networks and services.

By taking these steps, the Deutsche Post and the East German Ministry of Telecommunications had fundamentally decided the future development of their network. West German technology would be introduced to connect the two networks (as was foreseen in the plans for the overlay network) with the eastern network operator having gone further. The technology would be used to improve East Germany's network, since the goal of modernization was now to copy the West German model in all networks and services. Due to scarce resources, a complete modernization of the eastern network could not be foreseen.

The Deutsche Post's and the eastern telecommunications ministry's decision to cooperate closely with West German actors and build the overlay network can be interpreted in terms of a satisficing strategy.[16] The support put the Deutsche Post in a very satisfactory position compared to its past situation. It could draw on western technological expertise in the working groups of the "Joint Government Commission." It had access to digital technology since it now had hard currency at its disposal. There might have been better solutions but to find them would have required an active search. Since an attractive solution was offered to them, neither the Deutsche Post nor its ministry even tried to search for an alternative to West German support.

To conclude, the Deutsche Post had assigned priority to the expansion and modernization of its long-distance network rather than to a large increase in main lines. Even more importantly, the Deutsche Post had committed itself to the goal of compatibility with the West German network. There would still be two German telecommunications networks, but in the future the eastern network would become a copy of the western one.

By the end of this first period, East and West German actors had established cooperative relationships. This is mirrored by the network development plans, which aimed to interconnect the networks. In the future, they were to be fully compatible — but still separate.

March 1990 to June 1990:
Organizational Fusion and Technical Homogeneity

When the parties promoting unification won the GDR "Volkskammer" elections on March 18, 1990, it was evident that unification of the two states would take place within the next two years. This was a fundamental change in the actors' frame of reference. Of course, there had been intense speculations on the possibility of unification before. But now that the GDR government was also going to promote unification, all actors were forced to prepare themselves for the change.

The socialist government, which had maintained a certain distance to the West, had lost its influence after the elections. Immediately, leading Deutsche Post engineers seized what they considered to be the unique opportunity to modernize their network by cooperating closer with the Deutsche Bundespost. Since the preservation of the Deutsche Post's technical base was in jeopardy, renewal and expansion of its network could not be paid from the GDR means alone. It was likely that an independent but financially weak Deutsche Post could not ensure jobs for its employees in the future. Therefore, the engineers wanted to secure continued financial and technical aid from the Deutsche Bundespost. Material and technical goals now had priority over considerations about the organization's future.

In West Germany, it was undisputed that a modern telecommunication infrastructure was necessary for the economic development of a unified East Germany. One option that had been discussed was to use private capital to build up the eastern infrastructure, either by granting licenses for the construction of networks or by issuing Deutsche Post shares. Another option would have been joint ventures of international network operators with the Deutsche Post. Each of these options would have disrupted the political order of the FRG's telecommunication system after a unification because they would have established a private network operator alongside the state monopoly.

If DBP Telekom was to prevent this, it would have to take on the job of financing modernization in East Germany. Since DBP Telekom would not follow the models outlined above, it could only offer loans to the Deutsche Post. It was evident, however, that the Deutsche Post would not be able to pay back these loans. A merger with the Deutsche Bundespost would, in fact, transform the loans into an internal investment of the Deutsche Bundespost. Because DBP Telekom's monopoly rights were based upon the argument that only a monopoly could guarantee an equally high level of telecommunication infrastructure in all parts of the country, a low level of investments in East Germany would have endangered its position. This being the case, high investments in East Germany and an organizational merger with the Deutsche Post seemed to be necessary for DBP Telekom if it wanted to stabilize its monopoly position in a future united Germany.

DBP Telekom acted quickly, as others could possibly invest in the GDR. In the first week of April 1990, a committee known as a "management union" was established for the coordination of the Deutsche Bundespost and Deutsche Post policies. It decided on two goals. The first was to prepare for an organizational fusion of the two German Post and Telecommunication organizations. The second goal was to raise GDR telecommunications to the level of the Federal Republic. Working groups were set up to develop the modernization program, which was expected to cost DM 30 billion.

DBP Telekom management stated that such investments would be much easier to make in a united Telekom company. But the representatives of the East German ministry set priority on network improvements and the modernization program. Changes in the Deutsche Post organizational structure were to be made only in conjunction with the requirements of the modernization program.

The Federal Republic's telecommunications ministry took the view that there was no reasonable alternative to a merger of the two post and telecommunications organizations. An agreement on large investments in GDR telecommunications was also to include changes in Deutsche Post's organizational structures and a cutback of the Deutsche Post monopoly rights to the limits of European Community regulations. Obviously, the Deutsche Post and its ministry had to adopt the West German regulatory and organizational structures in exchange for support by the Deutsche Bundespost. On May 18, 1990, the East German Minister of Postal Services and Telecommunications formally agreed to work toward a uniform post and telecommunication system throughout Germany.[17] Thereby, he also followed the rapidly developing process of German unification. On the same day, the treaty on the German monetary union was signed.

A little later, by the end of June, the modernization program could be presented to the public. The merger of Deutsche Post and Deutsche Bundespost would take place at the same time as unification, which was expected to take place at the end of 1991.

The goal of the modernization program (called Telekom 2000) is to reach the West German level in the provision of telecommunication networks and services by 1997. To attain the West German phone distribution rate, the number of connections would be increased from the current 1.8 million to 9 million. This would be accomplished by connecting more subscribers each year. Starting with an additional 100,000 main lines in 1990, the Telekom planners intended to increase the number of new lines from 300,000 in 1991 to 1.5 million in 1997. The amount of investments was to grow correspondingly, starting with DM 1.3 billion in 1990, DM 4.5 billion in 1991 and up to DM 9 billion in 1997. Also, all DBP Telekom services such as data transmissions, mobile phones or cable tv would be introduced simultaneously. Investments for the Telekom 2000 program would total DM 55 billion.[18]

To prevent poor investments, every measure for improvement of East-West traffic was to adhere to the long-term conception of the network. To minimize costs, DBP Telekom has maintained the principle of minimization of technical heterogeneity in its network for decades. Its goal was to merge the networks and to set up one technically homogeneous network throughout Germany. By stipulating that only identical technologies be used in East and West, the planners also intended to prevent a hasty patchwork network expansion using a variety of technologies. Such an expansion would have brought rapid progress and facilitation to the users but would have cost the network operator additional expenses in the long run.

The expansion of the East German network was to follow a top-down overlay strategy according to which the long-distance network would be digitalized and integrated into the western network first.[19] A large increase in connections was projected to follow when digitalization had reached the local exchanges. Since the transformation of the long-distance network had priority over the expansion of the local loop, a rapid change in network structure could be obtained. From the DBP Telekom's point of view, the advantage of this strategy was the fast integration of the eastern network. By integrating the eastern network and creating a homogeneous and united network, DBP Telekom would gain control over the development of the East German infrastructure.

The Telekom 2000 projections put priority on maximizing the uniformity and homogeneity of the two German telephone networks rather than on rapid progress in connecting East and West or increasing the number of East German subscribers. Thus, the Telekom 2000 transformation strategy deliberately neglected present-day needs in its priorities.

Since only technologies used by DBP Telekom were to be introduced in the eastern network, DBP Telekom suppliers almost automatically had a hold on the new market. By June 1990, all of the larger West German telecommunication companies had found partners in the GDR, with whom they cooperated or were about to start a joint venture. There is little information on the situation of the numerous firms formerly belonging to the Kombinat Nachrichtenelektronik in the summer of 1990 since they had been disintegrated and partly reintegrated in renamed or newly founded companies. Considering that almost all production sites of the Kombinat Nachrichtenelektronik were taken over by West German manufacturers by the end of 1991 and that these takeovers proceeded the cooperations established then, nearly all parts of the former Kombinat that were still involved in telecommunications must have had western partners.

In a unifying Germany, the eastern firms had no alternative to a merger with western companies; their outdated products could not compete on western markets. When the monetary union was to come into effect by July 1, 1990, they would lose the advantage of low prices. It was very doubtful that they would keep their former customers, especially the Soviet Union, if they had to sell their products for West German currency.

Although the West German manufacturers had been motivated by the prospects of a high demand induced by the Telekom 2000 program, they ultimately invested for strategic reasons. Since they could possibly have satisfied the East German demand with the production at their West German plants, close cooperation with and investments in East German companies would keep new foreign competitors from entering the market during a boom.

This second phase is characterized by a change in cooperative relationships. Even though it had not happened by July 1990, it had become clear that West German actors would take over their eastern counterparts. The takeovers had a common motive. They were all intended to avoid the emergence of new East German actors as well as to keep foreign actors from entering the German telecommunications market. Apparently, unification with the GDR was not intended to disturb the constellation of actors in West German telecommunications but to extend it over united Germany.

This is more than the accession of the GDR to the Federal Republic would have required. It could have allowed two separate state telecommunication companies, as is still the case in railways.[20] Since the Telekom 2000 program had established the goal of a unitarian, homogeneous network using West German technology throughout Germany, i.e. the complete integration of the GDR telecommunication network, it also implicitly stipulated that only West German actors would take part in the modernization and integration of the East German telecommunications system.

July 1990 to December 1990:
The System Under Pressure Speeds Up

As of July 1990, there had been no improvements visible to users since November 1989, neither in the East-West bottleneck nor in the availability of lines in East Germany. Though there had been an increase in lines between East and West, this showed almost no effect. Being criticized for not improving the situation, Telekom said its efforts had always been overcome by the resulting increases in traffic. Deutsche Post had connected even fewer new subscribers than in the previous years. Trade and industry officials, struggling with a serious inability to communicate, said Telekom was prohibiting provisional solutions that would bring rapid improvements. DBP Telekom planners argued that provisional measures would not fit into their transformation program and would be too expensive. But "quick and dirty" methods were the antithesis of Telekom's engineering culture and working rules, which called for a solid, systematic network development based upon careful, long-range projections.

Since satellite links to the GDR could have helped to improve the situation, the Federal Republic's minister proposed permitting private telephone links to the GDR via satellite. The proposal was already very restrictive and excluded the connection of such links to the public telephone network. Its implementation was limited even further by the opposition of East and West Telekom as well as the eastern ministries and others who feared infringements on the state monopoly rights.[21] As a consequence, DBP Telekom did start to offer satellite services, but these had little effect due to very high fees.

In their negotiations on the future organizational structures in eastern Germany, Deutsche Post and Deutsche Bundespost management held different views. Deutsche Post officials intended to limit changes in personnel and proposed a gradual reorganization. In contrast, DBP Telekom managers favored a rigid tightening of the organization, immediate and complete restructuring, and full integration of the Deutsche Post into the Deutsche Bundespost branches. Since the DBP Telekom management considered Deutsche Post structures unfit for the technical transformation program, it expected identical structures to ease the control and guidance of the East German organization.[22] But in September 1990, the Deutsche Post still existed in its old organizational structure. Only the Deutsche Post offices had internally been split into telecommunications and postal services departments.

On the day of unification, October 3, 1990, the Deutsche Post formally became part of the Deutsche Bundespost and control was handed over to the Deutsche Bundespost management. Directly afterwards, the complete restructuring of the Deutsche Post began. The dissolution and recombination of offices, reshaping of districts and changes in responsibilities as well as the introduction of new working rules dispelled all possible tendencies towards inertia in the eastern telecommunication organization. It can be said that this restructuring broke the momentum of the telecommunication system of the GDR.

As critics had expected, the organizational restructuring, which took about half a year, interfered heavily with the network expansion program.

After unification, the performance of the telecommunications systems became more heavily criticized.[23] Industry representatives accused DBP Telekom, which was now responsible for the new states, of severely hindering economic development in the east by giving technical homogeneity of its network priority over customers' needs. Even the suspension of its monopoly rights within East Germany was being publicly discussed as a means to improve the situation. In addition, the Ministry of Finances even entertained thoughts of selling DBP Telekom to finance German unity. The united Telekom came under enormous pressure because the exceptional situation of unification permitted opinions to surface, calling for the most radical changes in the telecommunications system.

Since the terrible situation in telecommunications became an issue in the election campaigns for unified Germany's first national parliament, the government had to take action. Even though the telecommunications minister was in favor of further liberalization, he did not intend to overthrow the 1989 postal reform bill and start a fundamental reconfiguration of the Federal Republic's telecommunications system with a rapid introduction of private competitors to DBP Telekom.

By the end of October 1990, DBP Telekom finally reacted and sped up its program by adding DM one billion to its 1991 investments. But this didn't appease its critics. As the December elections drew near, chancellor Kohl intervened. In late November he ordered immediate steps to be taken, forcing DBP Telekom to speed up investments further. DBP Telekom now commissioned industry to set up complete networks for some parts of the former GDR, resulting in a projected 500,000 new East German subscriber lines in 1991.

It had been assumed that these so-called "turn-key" projects would lay the technical basis for the introduction of private network operators in East Germany, but this was not the case. DBP Telekom did not give up control over network construction since the projects were to become integral parts of the public network. Monopoly rights were not given up either. DBP Telekom had always commissioned private companies for the construction of exchanges or lines — but not for planning and supervision work. Coming to a make-or-buy decision, it delegated some projection and building supervision tasks to those companies which had already been in the building business with DBP Telekom.

Notably, delegation took place only within the circle of German telecommunication actors. Except for the acceleration of network expansion due to strong public pressure, DBP Telekom stuck to its program of integration and technical homogeneity. DBP Telekom's suppliers had gained a larger part of investments since Telekom had delegated some of its own tasks to the turn-key projects. On the other hand, delegation gave DBP Telekom the opportunity to continue with organizational restructuring.

After the turn-key decision, the debate on liberalization as a means to improve telecommunications with and within East Germany slowly ceased. But it was not until the summer of 1991, when the shortage in lines between East and West was overcome, that criticism against DBP Telekom stopped. Thereby the turn-key decision of early December 1990 set the final future structure and transformation strategy of the East German telecommunications system.

Summary

Using the projections of the future network structure as a guideline, the decision-making process behind the reconfiguration of the German telephone systems has been divided up into three periods. The end of each period is marked by a major decision on the technical configuration of the future network. The process covered the time from November 1989 to December 1990. During that time, the foundations of the future network structure and the transformation strategy were set.

The first period, from November 1989 to mid-March 1990, coincides with the phase of inter-German cooperation at the beginning of the unification process. East and West German network operators established a cooperative relationship. The transformation of the eastern system aimed at connection with the West, the long-term goal being full compatibility with the West German network.

The second period was one of rapid development, beginning in March 1990, when it became clear that unification would take place, and ending in June, shortly before the German monetary union went into effect. The network operators decided to merge their organizations in the future. At the end of June, a program for the transformation and integration of the eastern network was announced. Now, the goal was to construct a united, technically homogeneous network throughout Germany.

The third period started in July 1990 with the monetary union and ended shortly before the first national elections in unified Germany that December. The eastern actors as well as the network would be fully integrated into the western system. Since pressing needs of the day had been neglected, the system came under strong pressure. Telekom was forced to change its transformation strategy and expand the eastern network faster. By the end of this period, the future network structure and strategy were set.

What was decisive for the result of the reconfiguration process? State unification may have determined the telephone system's new configuration because the pace of the decision-making process followed the major events of state unification. Although the respective stages of unification had an influence on the options and strategies of the actors, the actors did not simply execute political orders. There had been room to maneuver, as in the question of if and how an organizational merger would take place or the issue of the way the systems would be technically merged. Furthermore, industry's behavior was not prescribed by the new institutional framework. State unification did provide the political framework for the unification process in telecommunications and strongly influenced its outcome, but it did not determine its final form.

The transformation process had been initiated by changes in the political environment and the changes in performance requirements emerging from them. Any transformation would have required an adaption to the new standards of performance, since, for instance, the new flows of traffic disturbed the functioning of the western system. The transformation described here cannot be understood as a process in which a system adapts by means of some internal mechanism to new requirements of its environment. It is quite the opposite since the system's actors developed a program that followed their interests but neglected the acute needs. Adaptation to the performance requirements of the environment had to be enforced by the threat of major interventions into the telephone system. The new configuration adapts to the new requirements, but it is not the result of a simple adaptation process.

Thus, this case underlines the importance of actors for the development of large technical systems. In this case, it was corporate actors who changed the system and not the individual "system builder" known from Hughes' work.[24] Not only did changes in the network projections follow the changes in the constellation of actors, they were also a mirror of the changing relationships between the actors. Cooperation leads to interconnection, merger to homogeneity. The technology and its configuration are not only a product of the situation but also a means to change it. A united and homogeneous network was intended to undermine the existence of two network operators. The limitation in the number of different technologies to be introduced also limited the number of actors involved.

The structure of the technical basis of the new German telephone system can therefore be understood as the result of the interaction among a set of corporate actors. Their interests and their acting shaped the technical configuration and the strategy to introduce this configuration.

List of Abbreviations

ANT	a West German telecommunications manufacturer
DBP Telekom	Deutsche Bundespost Telekom
FRG	Federal Republic of Germany
GDR	German Democratic Republic
ISDN	Integrated Services Digital Network
PKI	a West German telecommunications manufacturer
SED	Sozialistische Einheitspartei Deutschlands (Socialist party of the GDR)
SEL	a West German telecommunications manufacturer

Acknowledgements

I would like to thank Arne Kaijser, Renate Mayntz, Volker Schneider and Raymund Werle for helpful comments. Thanks also to Cynthia Lehmann for her friendly assistance.

Notes

1. For a typology of governance structures in telecommunciations, see Schneider 1991, pp. 23-26.

2. In 1989, 81 percent of annual surplus of the Deutsche Post (1.35 billion East German marks) went to the national budget. Günther and Uhlig 1992, p. 76.

3. The "national material balancing system" directed the distribution of goods in the East German economy. An assignment in the material balance sheet gave its holder the right to buy certain goods. Without such a claim, the Deutsche Post could obtain no telecommunications equipment. Günther and Uhlig 1992, pp. 57-85.

4. While the GDR's telephone distribution rate was comparable to countries such as Turkey, Poland or Argentina, these nations had a much lower GNP per capita. Tenzer and Uhlig 1991, p. 4.

5. An overview of the reform process is provided by Schmidt 1991.

6. Telekom 1991, p. 11. All other areas like data networks, terminal equipment or cellular radio were opened to competition.

7. Even though the Deutsche Bundespost branches are supposed to operate like private companies, they have kept the legal status of a public administration.

8. Its turnover is about two thirds of the turnover of Siemens or Volkswagen. The subsidies to the postal services and banking branches were DM 1.95 billion in 1990. Telekom 1991, p. 11.

9. These are Siemens, SEL, Alcatel, the Bosch Telecommunications branch (Bosch Telecom, ANT, Teldix and Telenorma), Philips Kommunikations Industrie (PKI) and DeTeWe.

10. On actors in West German telecommunciations see also Schmidt 1991; Schneider and Werle 1991.

11. ISDN is a service which makes it possible to connect different kinds of digital terminal equipment, e.g. computers and phones, to the same network access.

12. On the eastern side, the situation in East-West traffic was not as bad as on the western side. There were two reasons for this: (1) In the western network, there were 23,000 main extensions for every trunk line connecting to the GDR. Since the public network of the GDR was much smaller, the ratio there was 3,600 main extensions to one trunk line connecting to the West. (2) Most callers from the GDR could not dial directly but had to be connected by an operator. Thus, waiting queues could build up, which reduced the number of

attempts. In the fully automated western network, callers had to dial again and again, so one attempt to call to the GDR blocked the other.

13. As part of cold war politics, telephone traffic between East and West Germany was largely cut off in 1952. After being fully reopened in 1971, the number of lines between East and West increased only very slowly to 1,461 in November 1989. Tenzer and Uhlig 1991, p. 135.

14. The Postpauschale was the specific inter-German form of the annual payments by the Deutsche Bundespost to various foreign Post and Telecommunication administrations to allay the costs they incur handling calls, telegrams and mail originating from West Germany.

15. For years, the German-language Deutsche Bundespost publications had been the only source of information for the Deutsche Post engineers on the latest western technologies in telecommunications.

16. This is in contrast to a maximizing strategy. Simon 1964, pp. 257ff.

17. According to the Federal Republic's 1989 reform act, the Deutsche Post budget would be separated from the national budget. The Deutsche Post would be split up in three branches and its operative management taken out of the East German ministry.

18. The Telekom 2000 program in its final form is presented by Tenzer and Uhlig 1991.

19. The overlay strategy can be described as follows. Next to the existing analog network, a thin digital network is built, connecting only a few of the old network nodes first but growing in density. When the digital network has grown, the transition from analog to digital is done by means of gradual takeover of traffic.

20. The situation in railways is an exception in the unification process. Since the western railway was deep in debts and about to be reformed, its management refused a merger with the eastern railway company.

21. The Federal Republic's Ministry of Postal Services and Telecommunications began a procedure of licensing private telephone services via satellite in June 1990. The first licenses were not issued before the summer of 1991, when the East-West bottleneck had already been overcome.

22. Because the Federal Republic's Minister of Postal Services and Telecommunications guaranteed the takeover of almost all of DP's 130,000 employees, fear over the loss of jobs played only a minor role within the negotiations.

23. The East and West Telekom must have waited until unification to start their program. DBP Telekom did not invest DM 1.3 billion in 1990, as was first projected in the Telekom 2000 program. Data are not available, but statements range from DM 350 million to DM 690 million. Telekom 1991, p. 96.

24. Hughes 1987, pp. 54-64. The videotex cases also provide an example for corporate actors being system builders, see Mayntz and Schneider 1988, pp. 264-268.

References

This article is largely based on interviews with persons involved in the process, internal documents and newspaper publications. For reasons of brevity, these sources will not be referenced.

Auer, Eckart. 1991. "Aufbau der Fernmeldenetze in den neuen Bundesländern." *NTZ* 44(3):154-165.

Dingeldey, Ronald, Sabine Schulze and Wolf Kahle. 1991. "Der Weg zur Vereinigung der beiden Telekom-Unternehmen." *Telekom Praxis* 2:10-14.

Drescher, Joachim. 1992. *Die nachrichtentechnische Industrie in den neuen Bundesländern — ein Beispiel erfolgreicher Strukturanpassung.* Diskussionsbeiträge des Wissenschaftlichen Instituts für Kommunikationsdienste Nr. 82. Bad Honnef: WIK.

Günther, Wilfried and Heinz Uhlig. 1992. *Telekommunikation in der DDR. Die Entwicklung von 1945 — 1989.* Diskussionsbeiträge des Wissenschaftlichen Instituts für Kommunikationsdienste Nr. 90, Bad Honnef: WIK.

Hughes, Thomas P. 1987. "The Evolution of Large Technical Systems." In *The Social Construction of Technological Systems*, ed. Bijker, Wiebe E. et al. 51-82. Cambridge, Mass: MIT Press.

Mayntz, Renate. 1988. "Zur Entwicklung technischer Infrastruktursysteme." In *Differenzierung und Verselbständigung. Zur Entwicklung gesellschaftlicher Teilsysteme*, ed. Mayntz et al. 233-259. Frankfurt/Main: Campus.

Mayntz, Renate and Volker Schneider. 1988. "The Dynamics of System Development in a Comparative Perspective: Interactive Videotex in Germany, France and Britain." In *The Development of Large Technical Systems*, ed. Mayntz, R. and Hughes, T., 263-298. Frankfurt/Main: Campus.

Schmidt, Susanne K. 1991. "Taking the Long Road to Liberalization: Telecommunications Reform in the Federal Republic of Germany." *Telecommunications Policy* 15(3):209-222.

Schneider, Volker. 1991. "The Governance of Large Technical Systems: The Case of Telecommunications." In *Social Responses to Large Technical Systems: Control or Anticipation*, ed. LaPorte, Todd R., 19-42. Dordrecht: Kluver.

Schneider, Volker and Raymund Werle. 1991. "Policy Networks in the German Telecommunications Domain." In *Policy Networks: Empirical Evidence and Theoretical Considerations*, ed. Marin, Bernd and Renate Mayntz, 97-136. Frankfurt/Main: Campus.

Schnöring, Thomas. 1991. "Die Entwicklung des Telekommunikationssektors in den neuen Bundesländern." *Wissenschaftliches Institut für Kommunikationsdienste Newsletter* 2(2):24-27.

Simon, Herbert A. 1964. *Models of Man. Mathematical Essays on Rational Human Behavior in a Social Setting.* New York and London.

Telekom. 1991. *Das Jahr der Wende, ein Jahr des Aufbruchs. Bericht über das Geschäftsjahr 1990.* Bonn: DBP Telekom.

Tenzer, Gerd and Heinz Uhlig, ed. 1991a. *Telekom 2000. Telekommunikation für die neuen Bundesländer.* Heidelberg: R.v.Decker's.

Thiele, Klaus. 1991. "Telekom-Entwicklung in der ehemaligen DDR seit 1945." *NTZ* 44(3):150-153.

Werle, Raymund. 1990. *Telekommunikation in der Bundesrepublik Deutschland. Expansion, Differenzierung, Transformation.* Frankfurt/Main: Campus.

Witte, E., ed. 1990. *Telekommunikation in der DDR und der Bundesrepublik.* Heidelberg: R.v.Decker's.

7

The Australian Electric Power Industry and the Politics of Radical Reconfiguration

Stephen M. Salsbury

Structural Change in Australia's Economy

The structural transformation of the Australian economy, which began in 1974, has made it imperative to institute major changes in the production and distribution of electricity. From 1901, the year in which six British colonies federated to become the Commonwealth of Australia, to 1974 there existed in the new country a strong protectionist policy which built high tariff walls around a growing industrial economy. Australia is known abroad for its agricultural and mining sectors. The nation's farms are among the world's most efficient and Australia is a leading exporter of commodities such as wool, wheat, beef, sugar and cotton. Ever since the country's first gold rushes in the 1850s, the nation has been an exporter of gold, silver, diamonds, coal and a wide variety of base metals such as copper, lead, zinc and nickel.

Many North Americans, Asians and Europeans think of Australia as a nation of sheep stations, mining camps and wide, open spaces. In reality, approximately seventy percent of the nation's population lives in large metropolitan regions surrounding five great cities: Sydney, Melbourne, Adelaide, Perth and Brisbane. In these urban centers, large manufacturing complexes have developed that make basic commodities such as steel, copper wire and aluminium ingots, in addition to many consumer goods ranging from automobiles and refrigerators to radios and computers — indeed all of the manufactured commodities associated with a modern, industrialized economy. This is true despite Australia's small population of 17 million, which is slightly more than half the size of the American state of California with its 31 million inhabitants.

Australia imposed tariffs that were so high that it paid American and Japanese firms to establish large factories to manufacture automobiles which were Australian designed and made largely with Australian components. Until recently, the country had five major automobile manufacturing firms (two American and three Japanese).

All this began changing in 1974 when Australia began dismantling its tariff barriers and other restrictions on the free flow of goods in and out of the country. One of the reasons for the change in policy was the perception by the two major political parties Labor and Liberal (a conservative party) that the tariff propped up inefficient economic units. Australia is a country where organized labor is strong. From the British, Australia inherited craft unions which meant that any one firm, whether it be a railroad, an electricity supplier, or an automobile manufacturer, dealt with a large number of skill-based unions, such as electricians, plumbers, tool and die makers, and gas fitters. Such a union structure has made it difficult for management to reorganize factories and take maximum advantage of technological innovations when they are introduced.

Safe from international competition, Australian managers have normally passed on higher costs to consumers rather than confront the unions and work for radical restructuring within an organization. Union power has been so strong that most governments, whether Labor or Conservative, have found it too difficult to confront labor's power directly. Consequently in 1974 when a wage-driven hyper-inflation threatened Australia's economic stability, the then-Labor government under Gough Whitlam embarked on tariff reduction. The immediate goal was to control inflation by the admission of cheaper foreign goods. The long-term aim was to force management and unions to work together to make their industries competitive worldwide.

Economic reform began in the manufacturing establishments. However, the factory owners soon observed that it was not enough for reform to occur solely in their sector since many of their costs were determined by the services they were forced to use, such as the nation's port facilities, highways, railways and electric power systems. The manufacturing establishment insisted that it would be of no avail to reform their own activities without similar efforts in the country's infrastructure. Beginning in the 1980s, this perception set the stage for a thorough re-examination of the nation's state-owned electric power generation and distribution systems. Reformers have proposed the creation of a national electricity transmission grid where electricity would be "a tradable commodity." The aim is for an industry that would be nationally and internationally competitive and provide Australian consumers, especially the manufacturing sector, with the lowest possible price structure.[1]

This chapter will focus on the Electricity Commission of New South Wales' response to the challenge of restructuring Australia's economy. It should be emphasized that this chapter is not necessarily an endorsement of the recent history of Australia's electrical power industry. Rather, the aim is to analyze the unprecedented forces that are driving radical reconstruction. The politics of change are complex since it meant that the conservative labor government, first elected in 1983 and recently re-elected in 1993, has worked relatively easily with both socialist-labor and conservative state governments. The federal Labor government, especially under Prime Minister Paul Keating, dislikes state governments and to achieve reform wishes to centralize state planning and power in Canberra. The control of electricity through the operation of a national grid is part of this policy. Surprisingly enough, support for the national grid has even come from the Conservative New South Wales State government, partly because it emphasizes business competition and efficiency. Both Keating's Labor government and its conservative opposition at the national and state levels see lower power costs as essential to create jobs in the manufacturing sector. It is out of this curious ideological mix that radical reconstruction of Australia's power industry is proceeding.

Australia's Electric Power Industry

To understand the problem, one must appreciate Australia's geography and the history of the development of the nation's electric power generation and distribution systems. Australia is an island continent with a land mass approximately the size of the forty-eight continental United States (excluding Alaska and Hawaii). Mainland Australia contains five states (New South Wales, Victoria, Queensland, South Australia and Western Australia, see Figure 1). In addition, there are two territories: the Northern Territory and the small Australian Capital Territory. A sixth state, Tasmania, occupies an island about the size of the state of Oregon and is located immediately south of the southeastern portion of the continental land mass.

Although Australia occupies a large land area, most of the population is concentrated in the continent's southeastern portion. This includes southeastern Queensland (Brisbane), New South Wales (especially the coast, in particular the coastline 100 miles north and south of the great metropolis of Sydney), all of the state of Victoria

(Melbourne) and southeastern South Australia (Adelaide). The island state of Tasmania, which is not heavily populated, has rich hydropower resources and since it lies less than two hundred miles from Melbourne should be considered part of southeast Australia. A single state, Western Australia, occupies one-third of the Australian continent but metropolitan Perth is so remote that it is not easily integrated into the national economy.

Even though southeast Australia is a relatively small part of the continent (less than one-fourth of the land mass), the distances are great. Brisbane is nearly 600 miles north of Sydney. New South Wales' capital lies 600 miles north of Melbourne. Adelaide is about 400 miles west northwest of Melbourne. Furthermore, South Australia's capital is nearly 1,000 miles from Sydney and more than 1,200 miles from Queensland's capital, Brisbane. Southeastern Australia, therefore, is not one uniformly populated region but consists of four major metropolitan areas quite remote from one another. A fifth city and the nation's capital, Canberra is relatively small and located approximately 200 miles southwest of Sydney.

FIGURE 1 Australian States, Territories and Power Grid.

The Australian electric power industry primarily started to serve the capital cities of southeastern Australia. By the 1950s a common pattern had developed. The state of New South Wales is typical. In 1950 the state government created the Electricity Commission of New South Wales (ELCOM) which became responsible for the generation of electricity and power transmission throughout the state. The actual delivery of power to consumers has been accomplished by local government authorities, traditionally called County Councils, which purchase power from the Electricity Commission and distribute it to homes, businesses and factories in their territories. In a few areas, very large industries such as steel mills have made their own electricity, although this constitutes a minuscule portion of power generation in New South Wales. For all practical purposes ELCOM may be considered the only producer of electricity in the state with the exception of the Snowy Mountains Hydro-Electric Authority which will be considered later.

The State Electricity Commission of Victoria (SECV) provides the same service for that state as ELCOM. In South Australia, the producing and transmission organization is known as as the Electricity Trust, while in Queensland the corresponding organization is the Queensland Electricity Commission.

Australia's Coal Centers

Australia's three easternmost states of Queensland, New South Wales and Victoria all have large coal reserves near the major population centers. In fact, a great portion of New South Wales and Queensland are underlain by thick seams of bituminous coal suitable for either metallurgical purposes or power generation. Both states are massive coal exporters and coal is currently Australia's largest single export commodity. While Victoria lacks major bituminous deposits, it has vast reserves of brown coal in the La Trobe Valley, less than 100 miles from Melbourne. In addition, large oil and natural gas reserves exist off Victoria's coast in Bass Strait. South Australia's energy position is less favorable since its coal reserves are relatively small and are located several hundred miles from Adelaide. South Australia also contains important natural gas reserves in the Cooper Basin in the northern part of the state; these should be adequate for electricity generation until at least 1999.[2]

The formation of the Electricity Commission of New South Wales in 1950 was a landmark. At that time the state's power system was in chaos; several independent public authorities had been unable to meet demand, and prolonged strikes in the state's coal mines had disrupted the flow of fuel to power generating facilities. The strikes were only resolved when the federal Labor government under Ben Chifley sent in the Army — a move which broke the miner's union. (There was heavy irony in this since Chifley, an ardent unionist and committed socialist, had been fired as an engine driver during an equally bitter railway strike in 1917.)

The new Electricity Commission decided upon a strategy which emphasized the building of massive generating facilities on three major coal fields: the Western, which is 100 miles west across the Blue Mountains from Sydney; the Central Coast, approximately 80 miles north of Sydney; and the Hunter Valley, 130 miles north of Sydney (see Figure 1 p. 144). The commission's strategy was to build large generating units adjacent to coal mines which were owned and operated by the authority. In most cases, the state owned the mineral rights and found it easy and costless to transfer them to the authority. Building next to coal sources reduced transportation costs. The authority concurrently constructed high voltage transmission lines to the state's major centers of population, namely metropolitan Newcastle (at the mouth of the Hunter Valley), Sydney and Wollongong.

Placing generation units on the coalfields was revolutionary for New South Wales. Prior to the commission's founding, less than twenty percent of the power was generated by thermal stations which sat on coalfields. By 1958, forty percent of power generated came from the thermal stations located next to mine mouths and by 1982, ninety-five percent of New South Wales' power was generated by such stations.[3] The commission's policy was to build ever larger stations. The first 100 megawatt (MW) generator, Tallawarra, came on line in 1960 followed by the first 200 MW Vales Point plant in 1964, which in turn was followed by a 350 MW Munmorah unit in 1969.[4] After that date, the commission built several larger plants standardizing units which rated between 500 and 660 MW.[5]

For many years, ELCOM's policy seemed to pay off. The twenty-year period from 1958 to 1978 saw the average cost of the state's electricity fall by four percent per annum.[6] The commission attributed this to two main factors: the location of generation units on the coalfields and the economics of scale resulting from larger generating units. In 1958 the commission's largest unit had been rated at only 50 MW.

Power in New South Wales and Victoria

For most of ELCOM's history, New South Wales had been controlled by a Labor government which explicitly favored low electricity rates for households. Thus, the commission's tariff structure deliberately subsidized the consumer at the expense of the industrial and commercial users. This was partly possible because New South Wales, despite being Australia's most populous state, had little heavy manufacturing aside from its massive steel works which produced their own power. This picture changed in the 1970s when the Hunter Valley saw the rise of aluminium smelters.

In many respects, Victoria's power history parallels that of New South Wales. In the former case, the SECV built its large power generators on the brown coalfields. Like New South Wales, the SECV also produced its own coal and built a large, high-voltage transmission system to deliver electricity to the distribution networks. In one respect, though, Victoria differed from New South Wales. Melbourne is the nation's premier industrial city and the center of heavy manufacturing that includes automobiles, chemicals, machine tools and other power-hungry industries. In the 1970s, Victoria encouraged the development of an aluminium smelting industry at a site in Portland in the south-western portion of the state. This necessitated the stringing of heavy-duty, high-voltage transmission lines from metropolitan Melbourne westward almost to the South Australian border. While Labor governments dominated New South Wales until the 1980s, a conservative coalition ruled Victoria, favoring the delivery of low-cost power to encourage the state's industrial growth. Consequently, Victoria's tariffs provided the lowest cost industrial power of any Australian state generating authority.[7]

From the beginning, both Victoria and New South Wales planned their power systems independently of one another. Though similar in many respects, each state had its own philosophy for power generation. New South Wales emphasized low consumer prices as opposed to Victoria's goal of cheap power for industrial users. Furthermore, each state was fiercely protective of its own interests. For example, Australia has been late in its discovery and utilization of natural gas. Since the end of the Second World War, large fields have been discovered in Victoria along and under the Bass Strait, in the Cooper Basin of South Australia, in Queensland and on the Northwest Shelf of Western Australia. From the beginning, the states regarded natural gas as a scarce resource to be reserved for home use. Thus, Queensland does not permit gas to be sold to South Australia. In turn, South Australia has greatly restricted the export of gas from its fields to New South

Wales. In Victoria, which has massive quantities of this fuel, the state government gave a monopoly to its state-owned utility, the Gas and Fuel Corporation. Victorian government policy explicitly prohibits gas firing of base-load electricity generation. Under this policy, gas cannot be delivered to generators rated 500 MW or more. Gas generators can only be used for intermediate and peak load power plants and such power cannot be exported across state boundaries.[8]

New South Wales and Victoria might have gone their separate ways had not the federal government after the end of the Second World War constructed the Snowy Mountains Scheme. This large undertaking, comparable in many ways to the Tennessee Valley Authority in the United States, consisted of six large dams with massive storage capacity and the dual purpose of providing hydroelectricity and water for irrigation projects on the western slopes of Australia's Great Dividing Range. The scheme gets its name from the Snowy Mountains located in southeastern New South Wales.

The Snowy services seven hydropower stations. In 1990 the Snowy provided eleven percent of all electricity generation on Australia's mainland.[9] The Snowy is located far from any large city. After allowing for a small supply reserved for Australia's Capital Territory, two-thirds of the remaining power is sent to New South Wales and one-third to Victoria. To take advantage of this, the State Electricity Commissions of Victoria and New South Wales constructed transmission lines which linked their systems with the Snowy. Thus, from November 1959 the possibility existed for wheeling power from the coalfields of the Western, Central and Hunter regions of New South Wales south to Victoria by way of the transmission lines which served the Snowy.

The Snowy served another purpose as well. It was envisaged as off-peak and intermediate energy. It ideally complemented the large thermal units which became the rule in both Victoria and New South Wales. The Snowy was constructed so that water could be released to turn generators at peak times and then pumped back to the original reservoir by using base-load power during non-peak periods. The Snowy also made it possible for New South Wales and Victoria to trade their allotments as opportunity permitted.[10]

The Snowy Scheme made possible the beginnings of an electric power transmission grid which connected most of New South Wales' electrical generation system with that of Victoria's, as shown in Figure 1 (p. 144). Victoria's extension of its transmission lines to Portland provided further opportunity for extension of southeast Australia's grid and in December 1989, 275kV transmission lines were energized which connected Victoria with South Australia. In theory, this made possible the wheeling of power to Adelaide from the Hunter Valley through Victoria, a route distance of nearly 1,500 miles.[11]

Problems in New South Wales and Victoria, 1970-1990

The creation of an electrical grid that tied together three of Australia's four largest electrical generating and transmission systems was fortuitous. It did not involve central planning nor was it intended to undermine the authority and traditions of the three power systems. In fact, when the New South Wales and Victorian systems were first connected in 1959 it is fair to say that the officials of the systems of neither state recognized the opportunities. These only became fully recognized as the electricity systems in Victoria and New South Wales encountered serious difficulties in the late 1970s and early 1980s. After twenty years of apparent success between 1958 and 1978, New South Wales experienced an energy crisis. This was partly due to forces unleashed as a result of the Arab oil embargo and the rapid increase in energy prices during the early 1970s. Many in Australia saw this as an opportunity because the country was nearly self sufficient in oil and possessed massive and accessible coal and natural gas reserves.

As energy prices in the United States and other industrialized countries rose, companies that depended upon cheap power, especially aluminium firms, considered transferring their smelting activities from high-cost power countries such as the United States to low-cost producers such as Australia. The early 1970s saw New South Wales introduce four 500 MW generating units at Liddell which came on line between 1971 and 1973. This was entirely lucky since the lead time involved in building generating capacity meant that these units were planned before the oil shock. Chasing opportunities which were apparently beckoning, New South Wales built eight additional units of 660 MW each that came on line between 1984 and 1986. This expansion did not go smoothly. The Liddell Station was New South Wales' first use for open cut (as opposed to underground) coal. Having made this decision, ELCOM officials then changed policy and constructed the eight 660 MW units at the site of *underground* mines. The justification for this was to "avoid concentration of power projects in any one area of the state and to avoid undue dependence on inland water supplies or a single system of water supply for a power station for cooling purposes."[12]

In retrospect, these decisions proved disastrous for several reasons. First, the cost advantages associated with the new open-cut Liddell Power Station proved hard to capture. Despite Chifley's strike breaking, New South Wales continued to have powerful unions which were

— and still are — particularly entrenched in the Electricity Commission's mines and power stations. The unions were keenly aware of the importance of the Liddell plant to New South Wales' power system. In 1982, Liddell potentially accounted for twenty-four percent of power generated in New South Wales and the cost of production at that unit was fifty percent lower than the next most efficient unit. A large unit such as this provided a perfect opportunity for unions to use their muscle since a failure at Liddell caused maximum disruption to New South Wales' power system. A prolonged series of union disputes reached a peak in 1982 when Liddell's facilities often failed. The big power generators had an availability figure of around thirty percent as opposed to the availability factors of above seventy percent at two of the commission's other large plants and sixty percent at a third large generating unit.[13]

Secondly, while ELCOM was experiencing labor problems which nullified the advantages of Liddell, its decision to increase reliance on underground mines caused its fuel prices to increase. Open-cut mines are cheaper to operate since they minimize manpower. Furthermore, unions in the ELCOM mines ensured that underground miners received large wage increases that reinforced the disadvantage of underground as opposed to open-cut mines. For example, in 1980 the cost of underground mine coal was approximately twenty-five percent higher than open-cut coal. In 1982, this disparity widened so that underground coal was nearly forty percent more expensive than its open-cut counterpart. A change in policy, which directed that the very large units opened after Liddell used underground coal, had the result that the percentage of coal consumed by ELCOM coming from open-cut mines fell from 25.2 percent in 1980 to 13.7 percent in 1982.[14] The confluence of these forces led to the development that power costs, which had been falling consistently from 1958 to 1977, started to rise. In 1979 their increase was modest, but in the fiscal year of 1982 the cost of electricity in New South Wales rose in real terms by fourteen percent. This was followed by a further rise of twenty-nine percent in 1983.[15]

To make matters worse, the anticipated demand for energy in New South Wales did not match expectations. This slackening off was made abundantly clear by the reserve plant margin data. Reserve plant margin is the proportion of effective capacity in excess of peak demand. By 1989 New South Wales had one of the highest excess capacities in the world — forty-three percent as opposed to the Canadian average of nineteen percent and about the same for the Northeast Power Co-ordinating Council of the United States. A study by Australia's Bureau of Industry Economics found that only Tasmania, the Yukon in Canada and Canada's Northwest Territories had more excess capacity than New South Wales.[16]

The history of Victoria's Electricity Commission during the decades of the 1970s and 1980s has many parallels with New South Wales'. Victoria saw the world energy crisis of the 1970s as an opportunity. It launched a major program to increase generating capacity to cater for energy-intensive industries such as those for the conversion of bauxite into aluminium. In 1980, the state's installed generating capacity stood at 5,210 MW. An ambitious plan proposed to raise this generating capacity to 9,284 MW by 1989. The keystone was the construction of 2,000 MWs of capacity at Loy Yang A (four 500 MW units) and an additional 2,000 megawatts at Loy Yang B provided by four 500 MW units. Crucial to this goal was the completion of six of the 500 MW Loy Yang units by 1989 with the remaining two to come on line by 1991.[17] SECV management stressed the urgency of such construction since it not only involved expansion, but also the retirement of what it considered nearly 400 MWs of obsolete, inefficient power stations.

Victoria's Labor Unions

Victoria has a history, particularly in the public service sector, of militant, left wing, craft-based labor unions. The state's buoyant economy and the apparent urgency of massive new power generating capacity gave these unions an opportunity. The construction of a previous state large generator (Yallourn W) had already been subject to strikes that had caused construction to fall two years behind schedule.[18] Similar disputes immediately enveloped Loy Yang. The results have been catastrophic. The first of the B units scheduled for operation in 1988 is now rescheduled for 1993 and the second B unit for 1996.[19]

The high-voltage transmission lines connecting Victoria and New South Wales to the Snowy project blunted, however, the impact of the unions. New South Wales and Victoria had separate union organizations that did not coordinate their strikes. Therefore, in 1982 and 1990 when the New South Wales Liddell units failed due to labor trouble, the state imported much power from Victoria. Similar industrial problems at the SECV in 1989 and 1990 saw New South Wales send power southward. Because Victoria produced cheaper power than New South Wales, it became attractive for the latter to import power from the south to meet its peak load requirements rather than to start up relatively more expensive New South Wales generators.[20]

As in New South Wales, Victoria's economic climate changed dramatically between 1980 when the drive to increase the state's power generating capacity peaked and 1992 when the state was in the grip of a major depression that began in 1987 with the collapse of the Australian stock market. The decline of Victoria's economic activity together with the general downturn in the world economy lessened the need for power and forced a downward revision of power consumption estimates. Thus, the delay in the completion of the Loy Yang units proved less serious than might have been expected. In fact, Victoria's current state budget is in substantial deficit and its credit rating has been systematically downgraded by the international financial credit assessors Moody's and Standard & Poor's. This has made it progressively more difficult for Victoria to complete the rest of the Loy Yang complex. More importantly, the existence of substantial surplus power generating facilities in New South Wales gives Victoria the option to further delay the capital construction, instead importing power.[21]

Australia's Federal Government Supports Radical Reconfiguration

The Australian program to remove all tariff barriers, which has been accompanied by industrial contraction and prolonged eleven percent unemployment, together with the problems in the electricity commissions in Victoria and New South Wales, have caused the federal government to propose a radical reconfiguration of the nation's power industry. The federal government's Industry Commission concluded that "poor investment decisions" which resulted in "excess capacity and gross overstaffing" meant that electricity had not been supplied at the lowest cost. In addition, cross subsidization (with industry underwriting low electricity rates for householders), particularly in New South Wales and South Australia, penalized manufacturing and reduced job opportunities. The commission found that if Australia performed as well as international best practice and if cross subsidies between users were eliminated, the nation's "output would expand by $2.5 billion annually."[22] It also noted that some reform could be accomplished without radically altering the structure of Australia's electricity industry.

The Industry Commission looked with favor upon such "administrative reforms" as corporatism. Traditionally, state electricity commissions operated as government departments. Corporatism involves removing them from the mainstream of government and setting them up as separate units. This enables the electricity commissions to perform as private profit-oriented businesses. The goal is to distance such organizations from political expedience and enable them to re-examine such issues as cross subsidies, union feather bedding and community service obligations. Corporatized electric utilities should pay taxes (as do normal business enterprises) and be forced to meet rate-of-return targets on capital and pay dividends to their government owners.

The commission felt the corporate model would force governments to face explicitly the issue of providing low electricity rates for disadvantaged groups and to meet insuing costs not through cross subsidization but out of general tax revenue. The Industry Commission noted with approval that ELCOM had already been corporatized and had restructured its bulk tariff with the aim of ending cross subsidies. Nevertheless, the commission considered that "administrative change on its own will fail to lift the performance" of the electricity industry "to the fullest extent possible." This was because such reform did not address "the major factor underlying inefficiency — the lack of effective competition." The Industry Commission praised the New South Wales Government (now controlled by a conservative coalition) and stated that the "free market approach creates an atmosphere *more* likely to breed continuing improvement through ongoing market pressure for reform." Interestingly, the commission noted that the chairman of the SECV (appointed by a Socialist-Left Labor government) agreed, stating that "despite all we have done in SECV, feather-bedded manning agreements and restrictive management and work practices remain. From our experience, structural reform is necessary to shake them loose."[23]

The structural reform proposed by the Industry Commission received strong support from Australia's federal Labor government. The aim was to break up the electric industry by separating the three major functions of generation, transmission and distribution. Currently, as has been noted, the state power commissions effectively have a monopoly on generating and transmitting electric power. While links between the state networks exist, the high voltage power lines in each state are the exclusive property of that state's electricity commission. It is those commissions that determine the flow of electricity and set the prices for transmission. Under this system it has been difficult, if not impossible, for a large private generator of electric power, such as the Broken Hill Proprietary Company which makes power for its massive steel mills, to gain access to the electricity transmission lines. Nor has it been possible for a large consumer in New South Wales to make a contract to buy low-cost, brown coal-generated power from Victoria because New South Wales would refuse to wheel that power through its transmission system.

To introduce competition into Australia's electrical system, the Industry Commission has recommended, and the federal Labor government has agreed to set up, a federally controlled power transmission system which would take over all of the high voltage power transmission lines currently in place in New South Wales, Victoria and South Australia. Furthermore, the government would supply a minimum of $100 million to strengthen the interconnections between those states. The ultimate aim, however, would be to extend the national grid to Queensland and Tasmania. The latter is a major effort since it requires the construction of 200-mile-long underwater transmission cables beneath the Bass Strait which separates Victoria from Tasmania.[24]

The key to this system reform is to allow consumers — whether large industrial units or the County Councils which distribute electricity to households and small businesses — to purchase electricity from any source. In the words of the Industry Commission, the new

> transmission utility would be complemented by the development of market mechanisms which would supplant the traditional centralized planning approach. It would involve distributors negotiating forward supply arrangements with generators and with large users. Market trading mechanisms to allow distributors to sell power in excess of their contract requirements and, conversely, to purchase electricity to meet additional load would be required. The existing dispatch market whereby generators disclose their supply cost to the transmitter would need to be developed. This would allow for merit order dispatch of generating units so as to minimize supply cost for any given load. The transmitter would have to negotiate with generators to provide reserve capacity to ensure system integrity.[25]

In short, the national grid would allow for the creation of a nationwide market for electricity, putting pressure on the power generators to lower prices to maintain and utilize their capital plants. The Industry Commission argued that such structural change would enable the power generators to solve their over-staffing and cross-subsidization problems. In theory, under this system labor would be less likely to strike since lost production in one state could easily be replaced by production in another. Furthermore, a unit that gained a reputation for unreliability would face the prospect of losing power contracts and closing down.

The new restructuring has yet to take place. However, it has the strong support of the federal government and the New South Wales government. In July 1992, Victoria was ruled by a Labor government which owed a strong debt to left wing radical unions. Victoria's then premier, Joan Kirner, appeared to favor a national grid but faced the prospect that her party did not agree. The Socialist-Left faction (from which Premier Kirner came) and the power unions opposed moves which might lead toward privatization of state-owned industries. The Socialist-Left worked hard to stop privatization of the construction and operation of the unfinished Loy Yang B power generation units.[26] Kirner's Labor government was not united: in June 1992, Victoria's Finance Minister warned that "without reforms...the State Electricity Commission would become an 'industrial Stonehenge' swamped by states competing in the national power generation market".[27]

At the end of 1992, the conservative Liberal Party routed Labor in a bitter election. The new Kennett government, which is almost as radically conservative as the defeated Labor Kirner government was left wing, has no ideological problems with the privatization of Victoria's power generation. However, reform has stalled for other reasons, including Victoria's inability to borrow money, lack of power demand and conflict with Prime Minister Keating's federal Labor government. Basically, Victorian Liberals fear centralization of power in Canberra and are reluctant to lose control of valuable assets such as power generators and grids.

New South Wales Prepares for a National Grid

In the meantime, New South Wales is providing a model for the reconfiguration of the national system. The conservative Greiner government, first elected in 1988, inherited the problem of ELCOM's expansion phase in the late 1970s and early 1980s. The Greiner government appointed new Harvard Business School-trained managers for the system and asked them to solve the problems of excess capacity, overstaffing and operational inefficiency. The first problem was to send a signal to the commission's workforce that serious reconfiguration was going to take place. Hence, the commission closed smaller, uneconomic generating plants at three sites and the entire operation at the fourth site was decommissioned, including its coal mines. The latter caused a political storm which highlighted the loss of jobs; the commission's labor force declined from 10,600 in June 1988 to 6,700 in mid-1991.[28] It was significant that this reduction was accomplished without major strikes or power disruption.

ELCOM's strategy left it with six major power generation complexes in three regions: Western (Wallarawang), Hunter (Bayswater Power Station, Liddell Power Station) and Central Coast (Eraring Power Station, Vales Point Power Station and Munmorah Power Station). In addition, a seventh station (Mount Piper) is under construction in the West which will probably replace Wallerawang (see Table 1). The new configuration set up each complex, such as Bayswater, as a separate generating unit, the output and efficiency of which can be measured against a common yardstick.

The results have been clear both to workers and management. It will be recalled that in 1980 the Liddell unit was supposed to be the epitome of a large scale, efficient power generating unit. In the 1983 report of ELCOM entitled *Performance and Future Direction*, it was prophetized that "no power station built after Liddell . . . is likely to produce power more cheaply."[29] The labor troubles associated with Liddell proved that prediction incorrect. Nothing better illustrates this than the operating results for 1991 (Table 1). In 1991 Liddell, with 2,000 MW rated capacity, contributed 5,190 Gigawatt hours (GWh) or ten percent of the state's power requirements and had an availability of almost sixty-five percent. Liddell had a labor force of 588. This compared very unfavorably with Eraring, whose units had a rated capacity of 2,640 MW and produced almost twenty-six percent of the state's power (12,539 GWh) with an average availability of just over ninety percent and a workforce of 583. The state's other large unit, Bayswater, with a rated capacity of 2,640 MW produced almost thirty-two percent of the state's power (15,406 GWh) with an average availability of just over ninety-two percent and 587 employees.[30] The message is very clear: either Liddell improves or it will be closed.

TABLE 1 Major generation units in New South Wales July 1, 1990 to June 30, 1991.

Location	Generating Capacity (MW)	Production (GWH)	Percentage of Power Supplied	Availability	Employees
Hunter					
Bayswater	2,640	15,406	31.5	92.43	587
Liddell	2,000	5,190	10.6	64.62	588
Central Coast					
Eraring	2,640	12,539	25.6	90.41	583
Munmorah	1,200	3,658	7.5	72.70	497
Vales point	1,320	6,309	2.9	68.00	428
Western					
Wallerawang	1,000	2,750	5.6	75.35	534
Mount Piper	to replace Wallerawang	2,750	5.6	75.35	534

Source: ELCOM, *Annual Report,* 1991, pp. 33–34.

To introduce competition into New South Wales' system, the electricity commission has reformed its organization to consist of three separate power generating businesses, namely the Hunter, Central Coast and Western. Another major organization is the Power Transmission Grid, which proposes to behave in the same way as the proposed national grid. Under the new strategy, New South Wales power will be marketed through a "trading pool" where electricity is bought and sold. The pool will be run by the grid, which will determine the distribution of income to each of the pool's generating companies (Hunter, Central Coast, and Western). "The pool organization will meter the amount of electricity sold by the power stations to the pool through the competitive bidding process and the amount purchased from the pool by customers. It will then settle payments accordingly." The producing companies will be paid not only for the power they produce but also for their "reliability functions, such as plant availability, frequency control and 'back start' capability."

Under this regime, each power station will control its own income and expenditure, produce profit and loss statements and be required to make decisions, as would a normal business. The grid will also act as an independent enterprise. The grid will cover its costs by a published fee paid by both generators and customers for wheeling electricity from the point of generation to consumption.

While the grid is now available for use by the state's existing generators, the Snowy Mountains Scheme and the electricity commissions in Victoria and South Australia, it is envisioned that private generators will be able to wheel electricity over it for the same published charges that apply to current users.[31] In practice, it has not been easy for Pacific Power (the new name for ELCOM as the New South Wales' overriding authority that sets rates for its transmission lines) to fix competitive wheeling rates. Its traditional middle management (as opposed to the top management which the conservative Greiner government carefully selected to reflect its political views) still has a culture of vertical integration.

Australia's radical reconfiguration of its electrical industry is a bold attempt to free the enterprise from the rigidities of traditional, state-run economic ventures. It hopes to force down costs by putting the pressure of competition on labor and management. In many ways it comes at an opportune time. Organized labor's bargaining power has been weakened by Australia's worst economic depression since the Great Depression of the 1930s. Furthermore, the system is blessed with large overcapacity. The system also starts with a number of relatively small, inefficient generating units that can be easily shed — and which is already happening in New South Wales. For example, new state-of-the-art generating facilities at Mount Piper will soon replace the obsolescent generating facilities at Wallerawang. Similar opportunities exist in Victoria and South Australia, which will give the new reconfigured system an appearance of success in its early years.

Nevertheless, these opportunities will soon be exhausted and the economy is certain to rebound. This situation will create an environment in which the new network operates as a system of relatively tight capacity and buoyant employment opportunities. These were the conditions which made it possible for labor unions to disrupt the construction and operation of new facilities at Liddell and other New South Wales plants during the 1980s as well as at Loy Yang in Victoria.

The new network is based on two assumptions. First, that it will be more difficult for unions to disrupt a nation-wide power grid than state-based systems. This assumes disunity in the union movement, which has in fact characterized the 1970s and 1980s. Union disunity may no longer be taken for granted. Prime Minister Keating's "One Nation" concept also argues for union amalgamations into ever larger, more encompassing units. The current tendency in Australia's labor market is to create industry-wide unions which will replace the craft-based unions that now dominate the state electricity commissions. This creates the possibility that a single union might organize all power generating facilities in Australia. If this happens, unions might be able to thwart the competitive environment which the restructuring process hopes to create.

The second assumption behind the new network is that the restructuring process will open up for competition by allowing private enterprise to build power stations and wheel electricity over the national grid. There are relatively few large private power producers but it remains feasible for large industries, such as steel makers and aluminum smelters, to build power plants and sell excess power capacity to a national grid. New South Wales is also investigating the possibility of encouraging the construction of a purely private power station in the Upper Hunter region. Similarly, the new conservative government in Victoria might sell off the two unfinished Loy Yang power stations. For this to be totally effective, Victoria would have to grant private rights to mine brown coal in the La Trobe Valley. At present, competition through private power production is an unproven dream.

Conclusion

It should be emphasized that Australia's reason for reconfiguring its electric power industry is quite different from the motives for developing large transmission networks in the past. Thomas Hughes, in his path-breaking work *Networks of Power* indicates that the Pennsylvania-New Jersey interconnection, which in 1927 was "the world's largest integrated, centrally controlled pool of electric power (1.5 million kw)," concentrated primarily on seizing opportunities which arose because power peaks occurred at different times on the different interconnected systems. Hughes writes:

> A major benefit arose from load diversity. Throughout the year the Pennsylvania Power & Light Company had peak loads in the morning, for it supplied coal mines and other industries; the other two utilities, the 'seaboard' companies, experienced their largest loads in the evening. Furthermore, the Pennsylvania Power & Light yearly peak remained normally in the morning during October while for the other utilities the peak was in December in the evening.[32]

A classic example of the ability to capitalize on peak load differences occurs when an electrical grid connects two cities in different time zones or puts together a network connecting regions, one of which produces winter peak loads and the other summer peaks. An example of this was the recent merger of Oregon's winter-peaking Pacific Power with the summer-peaking Utah Power & Light. Other networks have attempted to take advantage of the different technical characteristics of electric systems such as marrying coal-fired thermal power systems with a system predominantly reliant upon hydropower. The advantages of such a merger are considerable. Thermal generation cannot be turned off and on at will; it takes many hours for a steam-powered unit to come on-line. Thermal systems typically use their most efficient plants to supply base load and fire-up their less efficient units to meet peak demands such as those caused by winter cold snaps or summer heat waves. In contrast, hydropower can be turned on instantly and, better still, hydropower usually costs less.

In the Australian case, the present system without *any* reconfiguration already maximizes the opportunities inherent in marrying hydro with thermal power. Currently, the Snowy scheme is used by both Victoria and New South Wales to meet peak-load demand. This allows Victoria and New South Wales to utilize their most efficient thermal units to supply base demand. For normal peak-loads, the two states do not need to keep less-efficient thermal units spinning since hydro is used to meet high demand periods which usually occur between 6:00

a.m. and 9:00 a.m. and between 4:00 p.m. and 7:00 p.m. Furthermore during slack periods, such as at night between 11:00 p.m. and 4:00 a.m. when base units have excess capacity, surplus power is used to pump scarce water from lower catchment resevoirs to higher dams so it can be used again. Both New South Wales and Victoria have already fully maximized the technical benefits of interconnection with the Snowy.

A large network which further strengthens the interconnection with South Australia and forms new links with Tasmania and Queensland will bring additional benefits of this kind. Tasmania, with its nearly 100 percent cheap hydro-powered system, enjoys capacity in excess of peak demand of 60.2 percent which could be well used to push down the cost of providing power to meet mainland peaks.[33] In addition, Tasmania is a winter-peaking state in contrast to Queensland and much of New South Wales, which peak in the summer. There are also eastwest benefits since Adelaide's morning and evening peaks occur forty-five minutes later than those in Sydney and Melbourne. State electricity commission managers in Australia understand all of these opportunities inherent in a large national network and would undoubtedly exploit the benefits as they have with the Snowy.

However, a radical reconstruction of Australia's networks has little to do with technical considerations or those factors which make a large-scale grid attractive by matching regions with different peak loads. Instead Australia's reconfiguration is designed to overcome political, social and economic problems. Australia's state-owned and -managed systems have been judged unable to confront entrenched labor organizations and have also fallen prey to the demands of their political masters, the state governments. The Australian solution is to break up the integrated electrical systems and shift the balance of power and management from the *generating organizations*, which have managed transmission lines in their own interests, to the *national grid system*. This solution is political rather than technical and the results will be watched with interest.

Abbreviations

AGPS	Australian Government Publications Service
AR	Annual Report
BIE	Bureau of Industry Economics
ELCOM	Electricity Commission of New South Wales
SECV	State Electricity Commission of Victoria

Notes

1. *Electricity Australia,* Journal of Electricity Supply Association of Australia, *Annual Report* 1991, p. 12.

2. The Electricity Trust of South Australia, *Annual Report 1989-1990,* p. 32.

3. *The Electricity Commission of New South Wales: Performance and Future Direction, Statutory Report to the Minister for Energy,* 1 February 1983, p. 22.

4. *Ibid.*

5. ELCOM, *Annual Report* 1991 p. 108.

6. Data adjusted by The Consumer Price Index.

7. *Industry Commission of Enquiry into Energy Generation and Distribution in Australia,* ELCOM 1990, Chapter 6: Pricing Policy, p. 49; see also BIE, *Research Report #40 International Performance Indicators Electricity,* Canberra: AGPS, February 1992, p. 33.

8. Industry Commission *Energy Generation and Distribution,* Vol I, *Summary and Draft Report* 15 January 1991, pp. 7-8.

9. BIE, *Research Report* #40, p. 40.

10. *Committee of Enquiry into Electricity Generation and Sharing of Power Resources in Southeast Australia* : Vol 1 Canberra: AGPS, 1982, p. 12-13.

11. Electricity Australia, *Annual Report* 1991, p. 9.

12. ELCOM, *Performance and Future Direction,* 1 February 1983, p. 27.

13. Ibid, pp. 43-48.

14. Ibid, pp. 64-65.

15. Ibid,p. 33.

16. BIE *Research Report* #40, p. 52.

17. SECV, *Annual Report* 1979-1980, p. 28.

18. Ibid., p. 29.

19. *Electricity Australia Annual Report* 991, p. 18.

20. *New South Wales Government Submission to the Industry Commission Enquiry into the Supply and Use of Electricity in Australia,* August 1990, Chapter B4, pp. 54-56.

21. Ibid.

22. Industry Commission, *Energy Generation and Distribution* 15 January 1991, Vol I pp. 1-2.

23. Ibid.,pp. 12-13.

24. *One Nation,* Statement by Prime Minister The Honourable P. J. Keating M.P., 26 February 1992, Canberra : AGPS 1992, p. 174.

25. Industry Commission, *Energy Generation and Distribution,* 15 January 1991, Vol I pp. 14-15.

26. *The Weekend Australian,* 23-24 May 1992.

27. Tony Sheehan cited in *The Australian,* 18 June 1992.

28. ELCOM, *Annual Report* 1991, p. 15.

29. ELCOM, *Performance and Future Direction,* 1 February 1983, p. 27.

30. ELCOM *Annual Report* 1991, p. 33.

31. The facts and quotations in this paragraph are taken from Murray Jackson "Pool Trading heralds a new era" in *Network*, a journal published by Pacific Power (formerly ELCOM), April 1992 Vol. 26, No. 2 , p. 10.

32. Thomas P. Hughes, *Networks of Power: Electrification in Western Society 1880-1930*, Baltimore: The John Hopkins University Press, 1983, pp. 331-333.

33. BIE, *Research Report #40*, p . 52.

References

The Australian, 1991.

Bureau of Industry Economics, *Research Report No. 40*.

Committee of Enquiry into Electricity Generation and Sharing of Power Resources in Southeast Australia, Volume One, Canberra: AGPS, 1982.

Electricity Australia, *Journal of Electricity Supply Association of Australia, Annual Report 1991*.

Electricity Commission of New South Wales, *Annual Report 1991*.

Electricity Commission of New South Wales, Industry Commission of Enquiry into Energy Generation and Distribution in Australia, 1990.

The Electricity Commission of New South Wales: Performance and Future Direction, Statutory Report to the Minister for Energy, 1 February 1983.

The Electricity Trust of South Australia, *Annual Report 1989-1990*.

Thomas P. Hughes, *Networks of Power: Electrification in Western Society 1880-1930*. Baltimore: The John Hopkins University Press, 1983.

Industry Commission, *Energy Generation and Distribution, Volume One: Summary and Draft Recommendations*, Draft Report 15 January 1991.

Network, a journal published by Pacific Power (formerly ELCOM), April 1992 Volume 26, No.2.

New South Wales Government Submission to the Industry Commission Enquiry into the Supply and Use of Electricity in Australia, August 1990,.

One Nation, Statement by Prime Minister The Honourable P. J. Keating M.P., 26 February 1992. Canberra: AGPS, 1992.

The State Electricity Commission of Victoria, *Annual Reports 1979-1980*.

8

Economics of Grid Systems in Reconfiguration: Competition in the Electricity Supply Industry

Olivier Coutard

Introduction

For decades, the electricity supply industry (ESI) was made up of large, *vertically integrated utilities* or interconnected, closely coordinated pools. Such monopolies generated, transmitted and distributed electric energy over wide areas to customers with no supply alternative. Independent power generation was marginal.

In recent years, however, this traditional structure has been increasingly challenged in industrialized countries. The introduction of competition in the ESI is today both a growing reality and a matter of sharp debate in the United States[1] and Europe.[2] In the U.S., this evolution admittedly was by far less radical, however, than that experienced by other *large technical systems* (LTSs) such as airlines, railways or telecommunications during the last decade of so-called deregulations.[3] In Europe, the initial ambition of the European Commission to promote an open market in electricity trade, and in particular to allow customers to freely choose their electricity suppliers, was confronted with determined opposition from industry[4] and many national governments.[5] But significant evolutions towards increased competition do now and will probably continue to occur in the ESI.

How these evolutions are made possible is complex to describe. As in the case of other LTSs, global studies of ESI evolutions must encompass six dimensions: technical and technological, economic, organizational, juridical and regulatory, political, and social dimensions.[6] We will see that both customers' and regulators' attitudes towards "network sectors" are essential in explaining current evolutions.

However, most arguments in the on-going regulatory debate are economic. Regulators, utilities, potential competitors for power generation, and researchers all focus on the economic features of electricity supply, economic conditions, and expected economic effects of contemplated changes. Although they share the same views on the goal to be pursued (namely, maximizing customers' benefits or welfare), they disagree on the appropriate way to reach it. Certainly global, economic evaluations will provide a useful measure of the benefits produced by regulatory changes. However, in this chapter, I will rather study the impact of economic analyses and decisions *on the process of change* itself.

Economic considerations are particularly complex and far-fetching in electricity supply systems because of their network or *grid-based organization*.[7] Grids inextricably mix technical, economic, organizational and political issues, so economic regulation regarding the ownership, planning, development, operation, access to, and use of grids must take into account wide-ranging effects.

This chapter is structured in three parts. In the first part, I analyze the traditional organization of electricity supply and emphasize the key function performed by interconnected networks in the traditional organization of electricity supply; this allows me to introduce useful concepts of power system operation and management. In the second part, based on a historical-juridical flashback, I describe how traditional electricity monopolies in the United States and Europe have been increasingly contested. I demonstrate that central coordination of grids is, however, preserved by all reformers seeking to increase competition in the ESI. The third part of the chapter discusses issues about the future organization of the grid raised by the new regulatory conception. Several examples from other large technical systems (postal system, telecommunications, and railways) prove helpful to understand regulatory evolution in the electricity sector. I examine how concepts developed from the experiences within these systems, specifically the "unbundling of services" and cost-based pricing, apply to the electricity sector.

Part One:
The Overshadowed Grid

In this first part, I briefly examine why electricity supply came to be considered as a natural monopoly in a given service territory. Strong management reasons were reinforced by legal and even political considerations. Interconnection networks or grids that link together all plants and loads are the core of electricity supply monopolies. However, for a variety of reasons, they were perceived by system managers and regulators as only a by-product of their supply activity with no economic value *per se*.

Natural Monopoly

All advanced economies have long had antitrust legislation. As far as democracies are concerned, U.S. Senator John Shermann summarized quite well the basic principle upon which antitrust legislation rests when he compared so-called trade autocrats' ability to prevent competition or to set prices to the power of kings or emperors.[8] Economic power, as well as political power, must be controlled and limited by citizens or governments.[9] Monopolies, more so than cartels or trusts, violate antitrust legislation: a monopolist has absolute power in a market.

From the point of view of social optimum, economic theory teaches that two sorts of monopolies have to be distinguished. *Natural monopolies*, on the one hand, are desirable as they represent the most efficient structure for a given economic activity; any market structure in which supply was provided by several independent actors would be less efficient. But natural monopolies call for regulation so that society is protected from their market power. On the other hand, *unnatural monopolies* have to be identified and broken up because they are both inefficient and dangerous.

Monopolies in general are easily identified, but distinguishing between natural and unnatural monopolies is another matter. A good example is provided by telecommunications. As Wenders notes, "there has been considerable disagreement, especially in the United States, over whether the telecommunications industry is a natural monopoly. The same data have been used to reach opposite conclusions and then to support or oppose regulatory intervention."[10] This observation could well apply to other sectors long considered to be natural monopolies.

To make things more complicated, the job of sorting out natural monopolies may not be left to the market alone. On the one hand, an unnatural monopoly can use its economic power to eliminate any potential competitor. Conversely, under specific conditions, a genuine *natural* monopoly may be unable to defend itself from unfair competition, i.e. competition contrary to the interests of customers. Regulation thus not only protects society from monopoly power but also protects natural monopolies from undue competition.

The concept of natural monopoly was first used by the U.S. gas supply industry[11] and soon re-used by some leading entrepreneurs in the ESI and, most prominently, by Samuel Insull (a leading Chicago entrepreneur in electricity supply who headed Commonwealth Edison Company for two decades). These actors argued that power supply would be best performed by monopolies and they admitted the necessity of — and even asked for — regulation of these monopolies to prevent abuse (monopolization).[12] This *quid pro quo* (regulation versus exemption from antitrust rules) was soon accepted by political authorities. Why were they so easily convinced?

Basic Notions About Power Systems

To answer this question, we will need to know a little about electrical engineering. It is a regrettable but well-known fact that electric power cannot be stored, at least not at an acceptable cost. Generation must therefore *instantaneously* balance power demand. A difference of a few percent between generation and demand causes a variation in electricity frequency, thus altering service quality. If this difference remains uncompensated for a minute or so, it may lead either to damages to the generator or customer equipment or even to local or more widespread blackouts.

The balance is, however, not easily maintained since both electrical lines and generation plants can be subject to unexpected outages. In order to face unexpected variations on the supply-side, operating and emergency generation reserves (i.e. additional power that can be obtained very quickly) have to be provided for. These reserves have a cost. In a one-plant system, the generation reserve necessary to face a plant outage is... another plant. In a two-plant interconnected system, the generation reserve is still only one plant (unless one wants to be protected against rare simultaneous outages). This shows how the reserve can be *shared* between interconnected plants. As plants grow larger for reasons of generation efficiency, reserve sharing becomes more and more economically and technically important.

In addition, individual demand varies widely between zero and a peak power demand for each customer. Providing generation capacity to satisfy all individual peak demands would be necessary in a one customer-one plant structure, but it is economical nonsense because of a *load factor* in individual consumption (customers do not all call for their peak demand all the time) partly reflected by a *diversity factor* (customers do not all call for their peak demand at the same time).

Note that the same holds true for transmission lines. Because of interconnection properties, lines within an interconnected grid are efficiently used. They are more loaded on average than a line that would connect a generation group with a consumption center (which is another effect of load and diversity factors). For a given total line length (or total cost), transmission service has a higher quality and is more reliable.

Finally, plants have very different operating costs. Interconnection allows *first* using the cheaper plants to meet demand in any part of the interconnected supply system (within the limits imposed on power flows by line capacity) and starting the most expensive plants only at peak hours: this is called *economic mix* (or *dispatch*). As a result, it is impossible to tell from which generation plant(s) the power consumed at a given node by a given customer comes. In other words, dispatch creates a total economic solidarity among grid users with total demand met by a single, interconnected generation system. Let us now turn back to the monopoly issue.

The Management of Interconnected Loads and Power Plants

Basically, a natural monopoly occurs when a single firm is able to meet the whole demand in a given market at a lower cost than any combination of several competing firms. But reducing costs and meeting demand are managerial tasks. Therefore the study of management principles ought to provide the key to the crucial monopoly issue: what exactly makes it more efficient to have one single supplier rather than several competitors? Thomas Hughes provides a useful starting point for answering this question when he summarizes as follows the principles acting as guidelines for achieving lower cost and greater reliability in electricity supply:

1. Obtaining economies of scale with large generating units, such as steam and water turbines;
2. Massing generating units near load centers or economical sources of energy and near cooling water at giant power plants;
3. Transmitting electricity to load centers through high voltage transmission lines;
4. Cultivating mass consumption by charging low and differential rates and allowing supply to create demand;

5. Interconnecting power plants to optimize their different characteristics;
6. Interconnecting loads to take advantage of diversity and thereby raising load and demand factors;
7. Centralizing control of interconnected loads and power plants by establishing dispatchings, or system coordinating centers;
8. Forecasting load requirements to achieve optimum operations within the interconnected system;
9. Lowering installed and reserve capacity and coordinating maintenance shutdowns through the exploitation of power plants interconnections;
10. Accepting government regulation to establish a natural monopoly;
11. Earning a regular and adequate return on investment to obtain capital at a reasonable interest.[13]

These principles, elaborated on in the 1920s, remained mostly unchanged for several decades. The traditional organization of the ESI was based upon them. The first rule was that low costs call for well-developed grids. Indeed, the importance of interconnection networks is obvious in this organization. The grid is necessary for the quest for optimal economic mix (principle 5) and maximal load factor (principle 6), both of which are identified by Hughes as the driving forces behind the growth of electricity systems.[14]

But the need for interconnection and distribution networks does not require that the grids and networks be monopoly owned. However, as can be seen in other large technical networks, the combination of legal constraints for infrastructure siting and anti-duplication concerns led to the emergence of franchise-based or state-owned monopolies over electrical lines. Airports, railway tracks and telephone lines are subject to the same sort of legal dispositions.

From Interconnection to Integration

The next step in system integration goes from monopolistic ownership of (or franchise over) electrical lines to supply monopolies. This shift was accompanied by three main sets of reasons.

First, there is what can be broadly described as *transaction cost considerations*. Because of the physical properties as described above, efficient and secure real-time operation involves load forecasts, advanced computer simulations and real-time *system-wide* monitoring. These demands led to the creation of system coordinating centers (Hughes' principle 7) where information could be gathered and computed at the lowest cost. It is important to note that such dispatching centers controlled and coordinated not only flows on grid lines but all power plant outputs at every instant. This is a consequence of the strong interactions created when plants are interconnected with the grid. Transaction cost concerns led to a monopoly in access to networks.

Secondly, in explaining the growth of supply monopolies, we must not underestimate the importance of so-called *public service requirements* in the integration process. The stronger the political will, the larger the integration; monopoly symbolizes equity. This especially holds true for services perceived as basic or essential, such as electricity, water supply or telephone service. And it is not just a symbol; a monopoly may be the easiest practical way to fulfill public service requirements, i.e. mainly requirements of universal service and rate averaging. A single firm is in a position to give all customers an equal share of the total cost of supply. Thus, public service considerations led to a monopoly in power supply, except for rare instances of self supply.

Thirdly, there is a logical consequence of the two points above. Competition in generation is discouraged by a legally enforced monopoly over access to grids and power supply, with independent power producers compelled to sell their output (in excess of their own consumption) to the local utility. As a result, independent power generation has remained marginal, thereby reinforcing the drive to monopoly organization.

The Overshadowed Grid

The process of grid integration and monopoly was fulfilled when electricity supply utilities had been made responsible for the development and maintenance of networks and generation plants, the operation of systems, and the supply of electricity to all customers.

The interconnected network was the backbone of this organization. Technically and economically, it provided the maximum benefits of interconnection; in terms of regulation, it was the core natural monopoly; in terms of ESI organization, it was the essential facility upon which supply monopolies could be built; and politically, it was a symbol of equity of access to the power supply.

However, the interconnected network was considered merely a by-product of the supply activity. Several reasons for this can be cited: (1) grid costs are low compared to generation and supply costs; (2) the concept of spatial economics and especially grid economics is complex, so its use by industry economists is less straightforward; (3) although dispersed by nature, grids were intended to homogenize service territories and thus tended to be idealized in a perfect interconnection function — what managers and regulators wanted it to be — rather than modelled as the dispersed systems which they are.

This sketchy panorama of the traditional ESI has given us tools to discuss the next question, namely what has been changing in the ESIs in the last two decades?

Part Two:
The Shrinking Monopoly

This part examines how first in the United States, then in Europe, traditional monopolies were increasingly contested by customers and/or regulators, who claimed there was a need for more diversity in the power generation industry. I will show how in this process, competition proponents recognized the need for a special regulation of grids. Yet they struck the heart of the ESI, namely the interconnection function. Many important issues raised by regulatory changes originated here.

Interutility Wheeling

First, it is important to note that the territorial monopolies in Europe and the U.S. were not isolated systems but most of the time interconnected within large, multi-utility systems. Exchanges between transmission-owning utilities soon developed for economical or security reasons. They mainly involved two contiguous utilities, one generating and the other consuming electricity. However, exchanges between non-contiguous utilities call for energy *wheeling* (i.e. the use of another utility's transmission assets).[15] These exchanges are significant since parties must rely on access to the grid of a third party, the wheeler, who is not directly interested in the exchange. Hence, the issue is whether or not to mandate wheeling.

In the U.S., interutility wheeling has been going on for decades. Such arrangements can be regarded as a first kind of competition in wholesale supply to utilities. However, the U.S. Congress has persistently rejected the notions of 'common carriage' or 'mandatory wheeling' on several occasions since the Federal Power Act in 1935, most recently in 1985.[16]

Interutility wheeling has also existed for a long time in Europe but it increased only recently when generation over-capacity in some countries met under-capacity in others.[17] (Traditionally countries sought energy self-sufficiency.) Such power exchanges have not always been easy to implement, with intermediate companies charging prices estimated to be too high. For example, a case recently arose concerning a contract between France and Portugal that would require wheeling over the Spanish grid.[18] A majority of European Community member states have supported a directive on transit.[19] This precisely requires the owners of high tension transmission grids to facilitate such transit between other transmission-owning utilities, if the transit is transnational and has no "opposite effects on quality and reliability."[20]

However, on both sides of the Atlantic, the "club" of transmission-owning utilities could have remained undisturbed. Starting in the 1970s, matters changed first in the U.S. and then in Europe.

Otter Tail Power Co. vs. the United States

Challenges to the U.S. ESI monopolies did not result primarily from the 1973 oil crisis, technical evolutions in the ESI itself or its environment (e.g. in computer applications for power systems), or radical advances in the economic theory of regulation. Instead they were initiated by a Supreme Court decision in 1973 in the *Otter Tail Power Co. vs. the U.S.* case. In the Otter Tail case, the Supreme Court held that *antitrust legislation* (Sherman Act, 1890) *was applicable to power companies.*

Otter Tail was a vertically integrated utility, meaning that it was an electricity generation, transmission and distribution company. It refused upon request either to sell power at wholesale prices or to wheel power from other suppliers to municipalities who wanted to establish their own municipal distribution systems. The court considered that Otter Tail's *refusal to wheel* power fell within the court's definition of *monopolization,* i.e. the possession of monopoly power in the relevant market and the abuse of that power. It ruled that Otter Tail should provide temporary transmission service to the plaintiff municipalities.

The Otter Tail case and the Supreme Court decision contain three important features. First, it is significant that the municipalities were *not* transmission-owning utilities. In this respect, they were comparable to any ordinary industrial customer and not a member of the "club." Second, in the view of the court, monopoly power does not result from vertical integration of ESI utilities but from grid control itself. This means that the court considers grid control — but not vertically integrated utilities — as a natural monopoly. It is reinforced by the decision itself: the Court would not have concluded there was an abuse of monopoly power in the refusal to wheel if it had been convinced that vertical monopolies were natural.

Thirdly, the court ruled that refusal to wheel had to be grounded on *technical* reasons and not, as could be expected, on economical ones. In this case, Otter Tail could not have argued merely that wheeling for the municipalities would have lowered global efficiency; it would be required to demonstrate that wheeling was technically impossible.

Public Utilities Regulatory Policy Act (PURPA)

The situation further evolved in the U.S. after the passage of the Public Utilities Regulatory Policy Act (PURPA) in 1978. PURPA allowed *non-utility generators* (i.e. electricity producers not regulated as public utilities) to supply electricity to utilities and required utilities to purchase this electricity. Price and contract conditions had to be specified by the individual states.

The passage of PURPA led to a considerable growth of non-utility generation.[21] Therefore, it tended to confront regulators not only with interutility wheeling but at the same time with *wholesale wheeling*.[22] Wholesale wheeling denotes transmitting electricity between a non-utility generator and a utility (to which it is not connected) which then sells the electricity to end-use consumers.

The growing importance of wheeling issues led the U.S. Federal Energy Regulatory Commission (FERC) to appoint a Transmission Task Force at the end of 1988 to address them. (Under the Federal Power Act of 1935, all transmission transactions on interstate grids are subject to the FERC's pricing jurisdiction; FERC's jurisdiction thus concerns grids in all states except the state-contained grids of Alaska, Hawaii and Texas.) The Task Force issued a report in October 1989 presenting how the commission's transmission policy could evolve.

Let us sum up the main innovations contained in the report. The task force expressed its concern about parochial behavior of transmission-owning utilities that favored the interests of the local customers, even when these interests were contradictory to general economic interest. To address this issue, the task force suggested that the authority of the FERC in grid matters be extended.

As Stalon puts it: "When an argument is made, as has been made, that a new transmission line is needed in Idaho, that the principal direct beneficiaries will be in Arizona and that both equity and efficiency say it should be paid for by utility customers in California, it is not likely that such a line will be built when needed, if at all, under the current regulatory system."[23]

The task force also considered the monopoly power of transmission-owning utilities. In the task force's view, this power has to be mitigated by giving access to third parties. However, up until 1992 (when the Energy Policy Act was passed, see below), Congress persistently rejected the notion of mandatory wheeling. Therefore, the task force suggested ruling on a case-by-case basis. This suggestion has been followed by the FERC on several recent occasions. In merger cases, the FERC required that merging transmission-owning utilities give third parties access to their grid so as to mitigate their (the merging utilities') market power (otherwise increased by the merger).

The Energy Policy Act[24] further enables the FERC to require a transmission-owning utility to provide wholesale wheeling services upon request, including wheeling between an independent power producer and a non-transmission owning distribution utility.

Towards the Opening of Grids

Mainly due to the increasing number of "non-club" members in the ESI, demands for *retail wheeling* have also appeared in recent years. These demands originate both from independent power producers and large industrial customers. Retail wheeling *is* considered a strategic issue. It occurs when a utility customer *bypasses* (leaves) its local utility in favor of another supplier such as a utility or a non-utility generator. The utility not only loses a customer but in addition must wheel power for the customer's benefit. The industry argues that bypass is not compatible with utilities' obligation to serve, while specifically raising the issue of *stranded investment*. Stranded investment refers to the fact that when investments have been made in order to supply all native load (stemming from this obligation to be served), if a customer leaves, the remaining fewer customers must share among them the bypasser's former share of investment costs. This is contrary to any criterium of fairness. Plummer summarizes the issue: "So it is an inherent characteristic of bypass that the bypassing customers will be 'winners,' and that the customers remaining behind on the regulated system will be 'losers.'"[25]

The overall efficiency of the electricity supply activity may still be increased by bypass if winners' gains exceed losers' losses. However, justifying and implementing bypass will not be easy, as it will raise protests by the losers, who would primarily be residential customers of the wheeling utility.

Indeed, retail wheeling is still marginal in the United States. Plummer lists only a dozen examples. But he notes: "It is the evolution here...that will be very strategically important over the next few years."[26] He further remarks: "It is important to realize that it is not something that awaits too much in the way of regulatory decisions."[27] The development of retail wheeling is mainly a "bottom-up" process.

The EC Treaties: Toward Third-Party Access

The shift in the United States' conception of regulated utilities from antitrust exemption to scrutiny and from supply monopoly to obligation to wheel can also be observed in Europe. Until the 1980s, European ESI monopolies were not challenged either by European institutions or by the national governments which regulated them. Since then, however, some European governments have reorganized their ESI. The United Kingdom's reform was by far the most radical, as it aimed at a complete *disintegration* of the previously vertically-integrated national monopoly utility (CEGB) and to the establishment of an electricity *spot market*. In this chapter, however, we will focus on changes at the European Community level.

In 1990, two important legislative measures presented by the European Commission[28] were adopted by the European Council:[29] the directive "concerning transparency of electricity prices charged to industrial end-users" and the directive "on transit of electricity on high voltage grids." It is worth noting that the U.S. Energy Policy Act of 1992 is more advanced in this respect, as it includes non-transmission owners as beneficiaries of mandatory wheeling.

However, a recent draft directive has outlined new, much more radical regulatory principles for the EC ESI. This draft directive "concerning common rules for the internal market in electricity" was prepared by the European Commission at the beginning of 1992. It requires member states to:

- allow undertakings established in the Community to build, operate, purchase or sell generating installations (art. 4, para. 1),
- grant licenses to build or operate electricity transmission or distribution lines (art. 5, para. 1), and
- ensure that any customer[30] established in their territory is able to purchase and to be supplied with electricity to be delivered from a producer or supplier in the same member state or in another member state" (art. 7, para. 2).

The 1990 EC directive on transit requires that wheeling prices be cost-oriented. The 1992 draft directive "concerning common rules... " contains dispositions on an open network provision in electricity supply. Its tenth preamble states: "Provision should be made for customers and producers of electricity to have access to the transmission and distribution systems without discrimination and subject to the availability of capacity and in return for reasonable remuneration." The terms "without discrimination" refer to the technical compatibility of connected equipment. As for "reasonable remuneration," this notion is developed in article 14 (para. 6): "The basis upon which the transmission operator's terms are set shall be such that the charges that would be payable are reasonably connected to the long-term costs incurred in the provision of the relevant service, together with a reasonable rate of return on capital employed in the provision of that service."

This draft directive is described by the commission as the second phase in a three-phase process of liberalization in the ESI. The first was the 1990 passage of directives on transit and on price transparency mentioned above and the third would "provide for further liberalization, including in particular a reduction in the barriers to the supply of electricity by producers to customers" (preamble no. 17).

The draft directive was presented but not adopted by the EC energy ministers' Council on November 30, 1992. However, "the Twelve said they were not demanding that Brussels withdraw or formally amend its proposals but rather 'modify' them."[31]

As in the U.S. case, this movement towards the opening-up of grids should not be seen primarily as the result of radical evolutions in the ESI or ESI environment. According to Hancher, it rather seems to be one of the consequences of the growing influence of community institutions after the adoption of the Single European Act (1985) and specifically of the European Commission. "Key industries such as energy, transport, water and telecommunications have escaped community action for over thirty years. The momentum generated by the drive to create a single market by 1992 has inevitably shaken this status quo."[32]

Hancher also emphasizes the role of national and European court rulings in this process. This is why this process may be described as a "top-down" one, i.e. one that was initiated at a top hierarchy — the European Commission — and intended to work its way down to the individual states' utilities.

The Shrinking Monopoly: Comparing U.S. and EC Approaches

Regulatory changes affecting the ESI in the U.S. and in the EC differ in many respects. In the United States, regulatory failure to cope with a changing environment, excess capacity and exceedingly high building costs of large generating plants (especially nuclear plants) created "bottom-up" pressures. The pressure for change came from independent power producers and large industrial customers. These pressures resulted in the "increasing role of competing suppliers of wholesale power to utilities for resale,"[33] a process which started with PURPA of 1978, which lead to more intense competition between power producers for access to markets and later to a partial opening up of networks in the Energy Policy Act of 1992.

In the EC, the evolution was more a "top-down" one that was initiated by the European Commission as its influence grew. Only afterwards did the initiative receive support from large industrial consumers (since there are virtually no independent power producers in Europe).[34]

However, the U.S. and European processes may well have a common origin. The times of ever-increasing demand for ever-cheaper basic power supply that led to the rise of uncontested monopolies (either by customers or public authorities) have come to an end.[35] Both processes resulted somehow in the same effects, although reforms in Europe have not attained the same level as in the United States. In EC jargon, monopoly in power generation is challenged as non-essential for the performance of a service of general economic interest. Monopoly control over access to grids is considered contrary to the principle that there should be no discrimination among member states according to nationality. The violation of the latter principle was all the more obvious since integrated systems have traditionally developed within national borders.

Thus, a common new conception underlies both processes. It can be summarized as follows:

1. Vertical integration of utilities is not justified by natural monopoly characteristics.
2. Therefore, independent power production must be allowed.
3. However, the development of far reaching, centrally coordinated grids induces large cost savings.
4. But grid control and coordination *per se* gives market power.
5. This market power must be mitigated for the sake of utility customers.

6. The development of independent power generation and the need to mitigate grid-based market power lead to forms of third-party access to grids.

It is interesting to note that both in the U.S. and in the EC, this new conception is fostered by regulatory institutions such as the FERC and the EC, which have authority over territories much wider than service territories of existing utilities. And at the same time as these regulators challenge monopolies, they encourage increased development of large, multi-utility grids.

Part Three:
The Valuable Network

The new regulatory conception raises a number of issues about the future organization of electricity grids. A crucial issue is defining sound relations between a single system *coordinator* that maximizes interconnection benefits and independent power *producers* and possibly also independent customers. New rules for the development, operation, and access to and use of grids have to be set.

In this part, I argue that it is mainly through these rules that a new balance between pure economic concerns and public service requirements can be reached. I will first turn to other network-structured large technical systems. Indeed sectors such as air transport, postal services, railways and telecommunications have been or are subject to analogous regulatory evolutions, after traditionally enjoying monopoly rights and exemption from antitrust scrutiny in exchange for regulation and public service requirements. Concepts such as the "unbundling (splitting) of services," the "open network provision"or "trans-European networks" were explicitly or implicitly transferred from these sectors. Therefore, analyzing changes in these other systems may help understand and even anticipate what is at stake in the ESI.

It is outside the scope of this chapter to detail even the most important EC initiatives touching upon the LTS regulatory environment. I will only present some aspects of regulatory evolutions in three of these sectors, namely postal services, telecommunications and railways. These sectors will illustrate the importance of network issues in the evolutions. Then I will turn back to the ESI to address network issues in more detail.

International Postal Services

National postal organizations have traditionally been forced to co-operate. For example, a letter sent to France from Germany is sorted in French sorting offices and delivered by French postal workers. Following international and mutual agreements on "terminal dues" within the Universal Postal Union (UPU), the German postal administration until recently remunerated this service by paying to the French administration twenty percent of postage income (depending on German mailing rates) arising from mail sent to France. The reverse applied in the opposite direction from France to Germany. Payments were then based on French postal rates.

The twenty percent share did not reflect real costs for the delivering administration. Final sorting and delivery of mail is the expensive part of the job — not initial national sorting performed by the issuing administration. But the agreement within the UPU rested on a principle of reciprocity between untroubled monopolies.

This situation is now questioned at the EC level. The European Commission has issued a so-called Green Paper favoring competition in postal services.[36] Increasing competition in postal services also affects relations between former postal administrations. In particular, large customers may take advantage of discrepancies among national postal rates. In principle, customers located in country A may be allowed to send their *inland* mail from country B (with lower postal rates), therefore depriving country A of considerable revenue. This is why former postal administrations oppose complete liberalization of international mail and have started renegotiating mutual agreements on terminal dues so that the delivery costs are better recovered.

Cost-based pricing requires "unbundling" or splitting an integrated postal service into separate basic collection, sorting and delivery postal services in a way that is accepted by other postal undertakings and regulators. This appears not to be straightforward.

Telecommunications

Another helpful example of the complexity and importance of network issues is provided by the telecommunications sector.[37] It is all the more interesting since the elementary notion of a "service of general economic interest" for infrastructures and vocal telephony has been expressed specifically in the 1987 Green Paper on Telecommunications.[38] It is thus interesting to study how common rules have been adapted to such an excluded sector.

From a historical perspective, deregulation in the telecommunications sector can be analyzed as the result of five mainly factors: inefficient regulations, emerging new telecommunications services, the development of computer-based activities, cross subsidies, and the emergence of a worldwide network. Here we are mainly interested in the two latter ones.

Telecommunication economists[39] have seen a positive network externality in traditional, voice telephony telecommunication networks whereby each subscriber benefits from the connection of other subcribers to the same network. However, some subscribers have higher incomes than others and are therefore ready to pay more for a given level of service. This induces cross-subsidies among groups of customers, from the high-income to low-income ones.

Cross-subsidization has occurred between undercost *access charges* and overcost *use charges* (especially trunk and international calls). This has resulted in cross-subsidies from industrial customers, who make a lot of long distance calls, to residential customers, who mainly make local calls. It is particularly obvious for traditional U.S. *flat rates* for local calls, but such cross-subsidization has been the rule in telephone networks throughout the world.

In its Green Paper on Telecommunications, the European Commission admitted that member states could maintain exclusive rights to operate network infrastructures and voice telephony (basic services), whereas all valued-added services should be opened-up to full competition. Concepts developed in the Green Paper were later implemented in European directives pertaining to the open network provision in the telecommunication sector[40] and to competition in the markets for telecommunications services.[41]

A difficult question the Green Paper raises is how to separate basic services from so-called value-added ones (another type of unbundling). (Examples of value-added services are the exchange of commercial information, videotex services and electronic data interchange.) U.S. regulators have failed to do so, and this has far-fetching consequences. In particular, public network infrastructure lines have to be leased to allow competition in the market for value-added services. The opening up of networks to new service providers allows these actors to provide large customers with low-cost connections and lower price voice telephony service. This is done by resale of leased line capacity for vocal telephony with no value added (what is called 'cream-skimming').

Competition eliminates subsidies from value-added services to basic services. Cream-skimming will further prevent subsidies from large to small customers of basic services and, to a lesser extent, from urban to rural customers. Yet this evolution has recently been given a strong push by the European Commission, which now seeks to liberalize vocal telephony as well.[42]

Railways

Let us turn last to railway companies. Traditionally, they were responsible for both railway system development and operation, but they are now subject to a directive[43] instituting the opening up of railway grids. Under this directive, national railways have to accommodate the running of third-party trains on their tracks.

This impacts on the traditional pricing procedures. As far as the French national railway company (SNCF) is concerned, infrastructure use pricing has up to now been based on the following differentiated principle:[44]

- Prices for goods' transportation are based on marginal costs, with state subsidies to compensate for unreconciled fixed costs. This is due to non-discrimination concerns. (The same principle applies to trucking, for which the "axle tax" does not include remuneration of road infrastructure fixed costs);
- Prices for passenger transportation are based on average or total costs, therefore rates ought to recover all costs. However, public policy requirements (such as reduced fares for the elderly, the disabled or large families) result in additional costs for the company which are also compensated for by state subsidies. In addition, the obligation to run unprofitable lines at averaged rates results in cross subsidies from profitable ones.

At least one pricing issue is obvious, namely whether the French government will be willing to subsidize goods transportation by non-SNCF trains on French railway lines. Other national railway companies are probably in similar situations, as reflected by the turmoil in member states which apparently followed the passage of the directive.

Similarly to the other examples from the postal and telecommunications sectors, access to and use of railway lines raise the issue of service unbundling. The management of infrastructure on the one hand and train-running on the other hand need to be dissociated in order to define non-discriminatory and fair access charges. Unbundling is incompatible with internal cross-subsidies since competitors may provide cheaper service on profitable lines and leave only the unprofitable ones for the former monopoly. This is another instance of "cream-skimming."

Competition and Network Issues

What conclusions may be drawn from these examples? In the new context, pricing procedures of connection to and use of network infrastructures are a strategic issue for former service monopolies, new service providers, public regulators and all customers. They have implications far beyond system economics.

First, as I examined in postal services, grid issues are a crucial dimension of the new relations among former "club" members. Secondly, imposing cost-based or cost-oriented pricing rules is not a panacea since unbundling in itself has a cost. Cost-based pricing can lead to vastly different prices depending on how costs are handled.

Thirdly, the new regulatory conception eventually challenges all cases and all forms of cross subsidies — from large industrial customers to small residential ones, between small customers, and between regulated and competitive activities in a given sector. This raises the issue of how to pay for universal service networks, traditionally financed by large cross subsidies, when service at cost is politically or socially unacceptable for specific groups of customers.

Learning from Other Systems: Cost-Based Pricing and Unbundling in the Electricity Sector

We have shown the departure of concepts such as cost-based pricing, unbundling of services or cross subsidies within European postal telecommunications and railway systems. Let us now examine how these concepts apply to the electricity sector. As noted above, the task force appointed by the FERC suggested that wheeling prices be cost-based, and the Energy Policy Act stipulates that wheeling prices must recover all costs induced by wheeling. In Europe, the EC directive on transit also requires that wheeling prices be "cost oriented." However, in traditional wheeling (i.e. wheeling between transmission-owning utilities), wheeling charges were not based on costs incurred by wheeling but rather on *split-the-savings* rules between the involved utilities, which the EC now rejects as potentially allowing the wheeling utility to abuse its market power.

On the other hand, grid costs are difficult to "unbundle" for two main reasons. First, generation and transmission are linked through the interconnection function, so it is not easy to distinguish between generation and grid costs. The dispatch process makes it impossible to tell which plant generates the power consumed by a given customer. Second, electricity supply — as well as other large technical networks — is characterized by large investments which benefit all grid users and operation costs incurred in common by grid users. There is no straightforward rule to optimally distribute these costs among users, and prices vary a lot according to the way costs are allocated.

The EC has a solution to the first issue, outlined in its draft directive: "the transmission and distribution functions of vertically integrated undertakings should be operated as separate divisions with separate accounts." However, this approach can be questioned, as argued for the U.S. telecommunications industry: "Requiring...regulated firms to create arms' length subsidiaries or divisions...certainly addresses the shared cost problem, but it is likely that the cure is worse than the disease."[45]

The unbundling of services also has a cost in terms of overall economic efficiency that may override benefits expected from increased access to grids, especially if this is done regardless of already ambiguous economic rules. The new ESI in the UK provides a good example of this. The rationale for short-term, integrated power system operation has been preserved and it is only in the long term that generation, transmission and distribution have been unbundled. The transmission business is still a regulated monopoly. Yet there is no evidence that the grid will develop in the long term so that the total cost of power supply is minimized, mainly because of difficulties in properly pricing transmission services.[46]

Public Service Requirements and Cross Subsidies

The "pure" economic problem just noted is complicated by political issues. In the examples from the other large technical systems, we observed how the unbundling of services led to the need for network use pricing and questioned the traditional regime of cross subsidies. What about the ESI?

I have described the strong economic solidarity created by the dispatch process in the traditional organization of the industry. Through dispatch, all generation costs are treated as common costs incurred by total electricity demand. I have examined how third-party access to grids challenges this conception.

But the unbundling of services also challenges economic solidarity in terms of network costs. For example, power supply in rural areas — although much more expensive than in dense urban areas because of high distribution costs — is in general charged the same rates for both subscription and consumption. Such rate averagings necessitate rather important subsidies since distribution costs can amount to a large share of total supply cost in some areas. These cross subsidies were allowed by the horizontal integration of rural and urban areas within large supply (vertical) monopolies.

So the grid can be viewed as interconnecting costs. It is important to note that the interconnection of generation to reduce total supply cost and the interconnection of rural and urban areas to average rates are two different things. Yet, both are characteristic of the traditional ESI and raise difficult political issues under current regulatory changes.

On the one hand, encouraged by the unbundling of services, large industrial customers seek to benefit from the cheapest generation available, leaving smaller residential customers with the more expensive ones or with stranded investment to pay for (when industries go to another supplier). They argue that their power needs are regular so they do not contribute to peaks.[47] This is analogous to cream-skimming in telecommunications. It modifies the respective contributions to total supply costs of large and small customers to favor large customers. Although it seems difficult to see evidence of the existence and volume of cross subsidies between these two groups of customers, it is worth keeping in mind that large customers asking for lower cost, stand-alone supply still want to benefit from network services, including guaranteed quality of service and security of supply which may be the supply of last resort. These network services also have a cost.

On the other hand, inter-area subsidies become impossible for unbundled distribution companies with separate accounts. This does question public service in areas where service at cost would result in prohibitive prices for some residential customers. Both changes tend to raise the cost of supply to some or all residential customers.

Conclusion

Competition is a growing reality in the ESI both in the United States and Europe. Other LTSs such as postal services, telecommunications or railways are experiencing similar regulatory evolutions. Increased competition in the *supply of services* is expected to enhance efficiency and customer benefits or welfare. Within the European Community, this evolution is reinforced by important dispositions of the Treaty of Rome on competition and the suppression of commercial discrimination based on nationality.

However, it is widely acknowledged that special regulations should be provided for grid infrastructure in order to manage *network externalities*, i.e. economic benefits that cannot be obtained through pure market mechanisms. These include general interconnection (allowing benefits from complementarities in supply and demand, quality of service, system robustness) and also risks associated with large technical systems. In the EC, benefits expected from well-developed networks gave rise to the dispositions on trans-European networks contained in the Maastricht Treaty.

The overall outcome of these evolutions depends mainly on the coordination between competitive service suppliers and a centrally-coordinated or strongly regulated grid. The basis for this coordination is grid use pricing but this issue is far more complex for a variety of reasons generated by network specifics.

From an economic theoretical point of view, grid use pricing is difficult because it involves spatial economics and because optimal pricing is much more complex in a competitive environment with many actors. Moreover, high fixed costs and important common operation costs that characterize grid-based large technical systems have to be distributed among consumers on little more than a rule-of-thumb basis. Cost-based pricing rules leave a lot of room for price discrimination, favoring some customers over others.

Practically speaking, networks exhibit other specifics that complicate pricing issues. They tend to ignore traditional political and regulatory borders so that ideal pricing rules may be inapplicable (see the Idaho line example noted earlier). Network infrastructures also involve crucial environmental aspects, exacerbating the NIMBY (not-in-my-backyard) syndrome. This can twist the otherwise homogeneous development of grids and impose additional costs to particular groups of customers. Finally, one has to deal with universal service requirements. With increasing cream-skimming and decreasing cross subsidies, other resources will have to be found to finance universal service.

Reformers of the ESI, therefore, face a number of challenges. As they elaborate and implement new economic regulations, and particularly those pertaining to grids, they are confronted with a myriad of far-reaching economic, organizational and political effects induced by these regulations.

Notes

1. Plummer & Troppmann, 1990 ; Simon, 1991.

2. Finon, 1990 ; McGowan, 1991; de Cockborne, 1991.

3. Joskow, 1989, p. 125.

4. Taccoen, 1991.

5. Finon, op. cit.

6. Joskow op. cit. & 1990; Hirsh, 1989; Finon, op. cit.; Bouttes & Lederer, 1990a.

7. "A grid-based system is a socio-technical system, often large, that is organized around a *specific distribution channel* that has been built solely for the purpose of delivering the system's specific commodity to its final destination (or providing its service to end users). The term 'grid' is figurative rather than literal." (Summerton, 1992, p. 75).

8. See Wilson, 1980.

9. Also, "Uncontrolled economic power is economically, politically and socially unacceptable in a democratic society." (Phillips, 1985).

10. Wenders, 1992.

11. By the American Gaslight Association as early as 1877, according to Platt, 1991.

12. Simon op. cit., pp. 88-92.

13. Hughes, 1983, pp. 370-371.

14. Op. cit., pp. 462-463.

15. "*Wheeling* is the use of one utility's transmission assets, perhaps complemented by changes in the pattern of use of its generation assets [in order to respect flow limits on grid lines], to transmit power as requested by another party." (Stalon, 1990, p. 6).

16. Henderson, 1990.

17. Persoz, 1991.

18. Finon, op. cit.

19. Council Directive 90/547 (1990).

20. Cardoso e Cunha, 1991; Hancher, 1991b.

21. Joskow, 1990.

22. Henderson, op. cit., p. 3.

23. Stalon, 1990, pp. 12-13.

24. Signed by U.S. President Bush on Oct. 24, 1992.

25. In Plummer and Troppmann, 1990, p. 48.

26. *Ibid.*, p. 8.

27. *Ibid.*, p. 4.

28. The EEC institution which concerns itself with initiatives to enforce legal dispositions contained in EEC treaties. Note that the commission is not a regulatory commission like the FERC. It is not entitled to set prices or pricing rules for service or goods supply. It may only expose violations of EEC treaties and require them to cease. This tool, however, is powerful enough as regards utilities.

29. In order to be enacted, most directives must meet with the approval of a qualified majority in the council.

30. Or, at least as a first step, any large customer.

31. *Power in Europe*, no. 138, December 4, 1992, p. 11.

32. Hancher, 1991a, pp. 256-257

33. Joskow, 1989.

34. Finon, op. cit. For the recent emergence of support to the Commission's initiatives, see for example Agency *Europe*, no. 5680, 92.03.02-03 for Belgian large consumers (FEBELIEC); no. 5684, March2-3, 1992 for the European chemical industry (including DOW, ICI, BASF, AZKO and the CEFIC). See also *Handelsblad*, Feb. 15, 1915 for the Dutch large consumers (SIGE).

35. I am endebted to Thomas Hughes for this point.

36. Green Paper on the Development of a Common Market for Postal Services, COM (91) 476 final (1992).

37. The discussion on telecommunications is based on Curien & Gensollen, 1992.

38. Green Paper on the Development of a Common Market for Telecommunications Services and Equipment, COM(87) 290 final (1987).

39. See for instance Allen, 1988.

40. Council Directive 90/387, July 24, 1990.

41. Council Directive 90/388, July 24, 1990.

42. *Le Monde*, March 17, 1993, p. 26.

43. Council Directive, JO L 237 25, Aug. 24, 1991.

44. Quinet, 1990.

45. Berg and Weismann, 1992, p. 454.

46. "Littlechild [the U.K. ESI regulator] has long been concerned that transmission charges did not give the right siting signals to new generators." (*Power in Europe*, no. 128, July 17, 1992, p. 4).

47. See the 1991 U.K. MEUC (Major Electricity Users Council) memorandum: "The price discrimination, which prevents users from negotiating (in the normal fashion in other markets) appropriate prices to reflect large regular demands, should be removed." (*Power in Europe*, no. 113, Dec. 5, 1991, pp. 6-7.

References

Baumol W.J., J.L. Panzar, and R.D. Willig. 1982. *Contestable markets and the theory of industry structure.* New York : Harcourt Brace Jovanovich.

Berg, Sanford V., and Dennis L. Weisman. 1992. "A guide to cross-subsidization and price predation: ten myths." *Telecommunications Policy* 16, 6: 447-459.

Bouttes, Jean-Paul. 1990. "Régulation technique et économique des réseaux électriques." *Flux — Cahiers scientifiques internationaux Réseaux et Territoires* 1. Paris: La documentation française.

Bouttes, Jean-Paul and Pierre Lederer. 1990a. "The organization of electricity systems and the behavior of players in Europe and the US". Paper presented at the conference *Organizing and Regulating Electric Utilities in the Nineties, a Euro-American Conference,* Paris, May 28-30.

_____. 1990b. "Towards a new industrial organization of the electricity sector in Europe ?" Paper presented at the conference *Organizing and Regulating Electric Utilities in the Nineties, a Euro-American Conference,* Paris, May 28-30.

Cahiers de l'IREPP. 1992. Special issue "les enjeux d'une économie postale." no. 12.

Curien, Nicolas. 1992. "Les caractéristiques techniques et économique des réseaux de télécommunications." Paper presented at the conference *Management des entreprises de réseaux,* Paris, January 20-21.

Curien, Nicolas and Michel Gensollen. 1990. "The opening-up of networks : planning or competition in the telecommunications industry and other public utilities." *Flux — Cahiers scientifiques internationaux Réseaux et Territoires* no. 1. Paris : La documentation française.

_____. 1992. *Economie des télécommunications, ouverture et réglementation.* Paris : ENSPTT-Economica.

De Cockborne, Jean-Eric. 1991 "La politique communautaire de la concurrence." *Annales des Mines*-série Réalités industrielles, numéro spécial 'l'Europe des grands réseaux'. (April).

Destival, Claude. 1991. "La déréglementation de l'électricité." *Enerpresse* 5439.

Driscoll, Dennis. 1990. "Access to the grid under European community law." Paper presented at the conference *Access and Pricing of Grid Systems.* Paris, November 15-17.

Finon, Dominique. 1990. "Opening access to European grids — In search of solid ground." *Energy policy.* (June).

Haag, Denis. 1992. "Déterminants techniques et économiques de la structure des systèmes électriques." Paper presented at the conference *Management des entreprises de réseaux,* Paris, 20-21 January.

Hancher, Leigh. 1991a. "European utilities policy — the emerging legal framework." *Utilities Policy* 1,3.

_____. 1991b. "EC survey — Recent developments affecting public utilities". *Utilities Policy* 1,3.

_____. 1992. "Competition and electricity markets." Paper presented at the conference *Integrated Electricity Market* (EET Ltd), Brussels, Jan. 27-29.

Henderson, Stephen J. 1990. "The evolving US transmission policy." Paper presented at the conference *Access and Pricing of Grid Systems,* Paris, November 15 -17.

Hirsh, Richard F. 1989. *Technology and transformation in the American electric utility industry.* Cambridge: Cambridge University Press.

Holmes, Andrew. 1990. *Electricity in Europe — Power and profit.* London : Financial Times business information.

Hughes, Thomas P. 1983. *Networks of power, electrification in Western societies (1880-1930).* Baltimore, Md.: Johns Hopkins University Press.

Joskow, Paul L. 1989. "Regulatory failure, regulatory reform and structural change in the electric power industry." *Brookings papers on economic activities: microeconomics*: 125-199.

Joskow, Paul. 1990. "The evolution of an independant power sector and competitive procurement of new generating capacity." Paper presented at the conference *Organizing and Regulating Electric Utilities in the Nineties, a Euro-American conference*, Paris, May 28-30.

Levy, D. 1990. "La tarification de l'électricité dans le monde, place de la tarification au coût marginal." *Economies et sociétés* 24,1. (January).

Maillart, Dominique. 1991. "L'accès des tiers aux réseaux électriques et gaziers en Europe". *Enerpresse* no. 5447. (November 13).

Mandil, Claude. 1991. "Les spécificités des réseaux énergétiques : la vision des pouvoirs publics français." *Annales des Mines*-série Réalités industrielles, numéro spécial 'l'Europe des grands réseaux'. (April).

McGowan, Francis. 1991. "Restructuring large technical systems : the debate over integration and deregulation of the EC electricity supply industry." Paper presented at the *Third international Conference on the Dynamics of Large Scale Technical Systems*, Sydney, July 1-6.

Persoz, Henri. 1986. "Le transport de l'énergie électrique." *Metropolis*. no. 73-74.

_____. 1991. "L'interconnexion des réseaux électriques : de l'Europe de l'ouest à l'Europe élargie". *Annales des Mines*-série Réalités industrielles, numéro spécial 'l'Europe des grands réseaux', (April).

Phillips, Charles F., Jr. 1985. *The regulation of public utilities. Theory and practise*. Arlington: Public utilities reports Inc.

Platt, Harold L. 1991. *The electric city 1880-1930. Energy and the growth of the Chicago area*. Chicago: Chicago University Press.

Plummer, James L. and Susan Troppmann, eds. 1990. *Competition in electricity : New markets and new structures*. Public Utilities Reports Inc. and QED Research Inc.

Schweppe, F.C., M.C. Caramanis, R.D. Tabors, and R.E. Bohn. 1988. *Spot pricing of electricity*. Boston: Kluwer academic publishers.

Simon, Jean Paul. 1991. *L'esprit des règles, réseaux et réglementation aux Etats-Unis*. Paris : L'Harmattan (Coll. 'Logiques juridiques').

Stalon, Charles G. 1990. "Current issues in transmission access and pricing in the US electric industry." Paper presented at the conference *Organizing and Regulating Electric Utilities in the Nineties, a Euro-American Conference*, Paris, May 28-30.

Summerton, Jane. 1992. *District heating comes to town, the social shaping of an energy system*. Linköping (Sweden): Linköping Studies in Arts and Science.

Summerton, Jane and Ted K. Bradshaw. 1991. "Towards a dispersed electrical system: challenges to the grid." *Energy Policy*. (Jan.-Feb.).

Taccoen, Lionel. 1991. "La commission des communautés européennes et le secteur électrique." *Annales des Mines*-série Réalités industrielles, numéro spécial 'l'Europe des grands réseaux'. (April).

Wenders, J. 1988. "Unnatural monopoly in telecommunications." *Telecommunications policy* 16,1.

Wilson, J. 1980. *The politics of regulation*. New York: Basic books.

Confronting Cultural Incompatibilities

PART THREE

Confronting Education Inequalities

9

The Internet Challenge: Conflict and Compromise in Computer Networking

Janet Abbate

Introduction

Standards battles are a useful site for historical inquiry into the development of large technological systems. The level of struggle between competing standards can provide an indicator of the economic importance of the technologies involved, as in the "Battle of the Systems" between AC and DC electrical generation.[1] Standards disputes can also reveal different actors' perceptions of the goals of a technological system and how these goals can best be achieved. The conflict that arose in the mid-1970s over computer networking standards provides a case study of the social and technical implications of standards debates.

In the early 1970s computer networks were growing in importance as research tools, strategic components of military command and control systems, and commercial enterprises. Research networks had been successfully established in the United States and Europe beginning in the late 1960s with the Mark I at England's National Physical Laboratory, America's ARPANET, and France's CYCLADES. Computer corporations were starting to offer proprietary network systems as an enhancement to their hardware and software products. The market leader, IBM, introduced its Systems Network Architecture in late 1974; Xerox, Digital Equipment Corporation, and other large manufacturers soon followed suit. In the United States, Telenet began building the first commercial data network in 1972, while in Europe, Canada and Japan national telecommunications carriers planned to add data transmission to their array of services. Many of these new systems were wide-area networks, which stretched beyond a single building or campus to connect distant sites.

As more regional and wide-area networks were built, people began to see opportunities to connect networks together, a process known as *internetworking*. American universities and European research centers wanted to link their campus networks to provide regional resources. The U.S. government owned several different types of networks and saw opportunities for combining their services. State and private telecommunications carriers wanted to expand into the market for data transmission. But though they shared the goal of connecting systems, these various groups represented radically different views, interests, traditions of practice, and constructions of networking, which led them to disagree on fundamental design decisions. When international standards for networking were proposed, they became the focus for extensive and often heated debate.

The conflict over competing networking standards reveals both conflicts of interest between the groups participating in computer networking and fundamental differences in their perceptions of the technology. I use a close reading of the standards debate to illuminate the often unspoken assumptions and concerns of these various interest groups. I then consider how the fate of network technologies depended on their ability to adapt to changing user demands and to the existence of other, incompatible systems.

How Networks Work

Wide-area networks can be divided into three functional areas. The first is the *communications subnet*, which moves data between users' computers (called *host* computers). The subnet is composed of transmission links (such as telephone lines, microwave, radio, or satellite links) connected by computers called *nodes* that switch incoming data from one link to the next. Subnets commonly employ a technique called *packet switching*, in which the flow of computer data is divided into small units called packets. Packets from many different connections can then be interleaved along the same transmission link to create a more constant and efficient flow of traffic. The interaction between host computers and the subnet forms a second functional area. The third area is end-to-end transactions: interactions between a pair of host computers which are conducted via the subnet (see Figure 1).

Activities in all three areas are regulated by *protocols*, which specify how data will be formatted and what procedures will be followed in any type of transaction. *Network* protocols determine how packets get from one side of the subnet to another. Host computers use *transport* protocols (optional in some networks), which add an extra layer of reliability and control to the basic network transmission service, and *application* protocols, which offer specific services such as file transfer, remote login, or electronic mail.

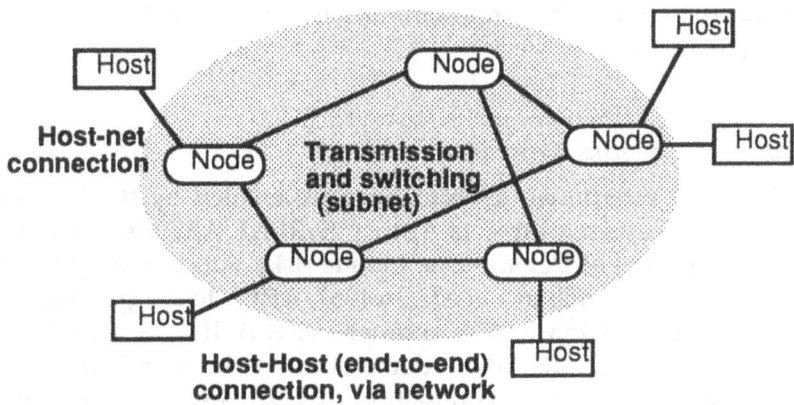

FIGURE 1 Network Functions.

All networks can be expected to lose or damage some of the packets they transmit. Network designers must therefore decide how to provide end-to-end reliability, the guarantee that a packet sent from one host will arrive without error at the destination host. They must also decide on a flow control mechanism, to keep the sender from overwhelming the subnet or the receiver with data. Discussions of data communications in the mid-1970s focused on two alternative techniques for providing these services. In a *virtual circuit* network, the subnet would keep track of each individual connection, setting up a path from node to node so that each connection would appear to have its own dedicated circuit. The subnet would provide flow control for each connection and guarantee the reliable delivery of packets. This would require considerable processing power in the subnet nodes; however, a transport protocol would not be needed, so less processing would be demanded from the hosts.

In a *datagram* network, the subnet would provide a more minimal service. Each packet would be sent through the subnet on its own, and different packets might take different paths to their destination. The subnet would not keep track of which packets belong to which conversations, instead leaving the hosts at either end of the connection to collect and sort out the packets. Hosts would be responsible for error correction and flow control. This meant less processing would be done in the subnet; however, a transport protocol would be required on the hosts to manage the connections.

Either technique could supply the end user with a reliable and seemingly dedicated connection. The difference was whether the responsibility for providing the end-to-end connection would be concentrated inside the subnet (a virtual circuit network) or in the hosts (a datagram network).

The ARPA Internet Program

The first internetwork protocols were created by the U.S. Defense Advanced Research Projects Agency (ARPA). ARPA's mission within the Department of Defense is to pursue risky but promising research in areas relevant to military needs. In 1967, ARPA had begun work on the ARPANET, a nation-wide computer network linking ARPA research contractors. The ARPANET demonstrated the feasibility of packet switching, and after its completion in 1972 ARPA began work on a packet radio network called PRNET and a packet satellite network called SATNET. In order to connect its various networking projects, ARPA began its Internet program in 1973.

Since ARPA was experimenting with different types of hardware and protocols, the Internet program was geared to connecting networks with very diverse characteristics. The ARPA team decided that their internet system should expect nothing more than datagram service from any network; while the protocols would run over virtual circuit networks as well, they would not depend on this higher level of service. The guarantee of reliability would be a host-based transport protocol, called the Transmission Control Protocol (TCP), which creates an orderly, error-free virtual circuit out of the stream of datagrams presented by the subnet.

Another belief guiding ARPA's effort was that users would want to connect not only individual computers but entire local networks across an internet. Since the user could not count on all of these networks using the same procedures, ARPA saw the need for an additional Internet Protocol (IP) to provide a uniform method of addressing among connected networks. Under the resulting system, any network that wants to join an internet based on ARPA's protocols must use IP; TCP is optional for internet connection but is required to support many applications. TCP and IP are known collectively as the TCP/IP protocol suite.

Led by ARPANET veterans Vinton Cerf and Robert Kahn, ARPA's Internet team produced the first version of its protocols in 1974. By 1976 ARPA had connected ARPANET, PRNET, and SATNET. The U.S. Department of Defense made TCP/IP its official network standard in 1980, and when the National Science Foundation began building a nation-wide research network in 1985 it also chose TCP/IP. ARPA made its protocol specifications freely available and promoted their use by funding implementations of TCP/IP for UNIX™, a widely-used operating system. Users building their own networks were attracted to TCP/IP by its successful use in the ARPANET, its easy availability, and the prospect of being able to exchange information with many U.S. research centers. It soon became the most popular non-proprietary protocol suite in the United States. The widespread use of TCP/IP made it possible for ARPA, the National Science Foundation, and private network owners to set up links between their systems for the exchange of information. The resulting collection of interconnected TCP/IP networks became known as the Internet.

Public Networks

A second networking effort was led by the Post, Telegraph and Telephone administrations (PTTs). The PTTs, who operate state-owned telephone networks, wanted to tap the growing market for computer services by providing common carrier networks for computer data, known as public data networks. In 1974 and 1975 carriers announced plans to provide public data networks in Canada, the United States, France, Britain, and across Europe. The PTTs realized that if they did not explicitly agree on techniques for building and connecting these networks, they would run the risk of creating incompatible systems.

The public carriers did not want to base their networks on proprietary protocols. They objected in principle to being dependent on protocols that could only be used with one manufacturer's equipment, particularly since computer manufacturing was dominated by American corporations while the public data network movement was led by national carriers in Europe, Canada and Japan. These countries saw non-proprietary protocols as a way to give their domestic manufacturers a better chance to compete with IBM, while putting the PTTs firmly in control of network design.

Canada, which was farthest along in building its public data network, was the first to reject IBM's network protocols. In October 1974, the Trans-Canada Telephone System had begun work on its Datapac packet-switched network. IBM computers made up a large proportion of the proposed network, but rather than use the IBM protocols, the Canadian government issued a statement in support of publicly-available protocols that would work with a variety of equipment. Trans-Canada tried to persuade IBM to adapt its protocols to Datapac's requirements; IBM argued that it did not make sense to modify its protocols to suit a small segment of its market. By mid-1975, the two sides were at a standoff.[2]

To resolve this dilemma, the carriers turned to the Consultative Committee on International Telegraphy and Telephony (CCITT), an international organization that researches and develops technical standards for telecommunications. CCITT membership is open to national telecommunications carriers; private carriers and industrial and scientific organizations may participate in an advisory capacity, but cannot vote.[3] CCITT publishes its standards in the form of non-binding recommendations. Representatives from some of the emerging public data networks decided to create a CCITT Recommendation that would establish a standard protocol for host-network interaction.

Some CCITT participants suggested adopting the TCP/IP protocols as a standard. This idea was rejected, however, possibly from a reluctance to adopt any American-made standards.[4] Instead, a team led by representatives from Bell Canada, the French PTT, the American network provider Telenet, and the British Post Office began work on a set of protocols called Recommendation X.25. They decided that the new protocol would specify a virtual circuit network, in which the subnet would be responsible for call set-up, flow control for each connection, reliable delivery of packets, and recovery from hardware failures in the host or network.[5]

At the CCITT's 1976 plenary X.25 was approved by a majority of voting members, and public data networks that had been in the planning or development stage when X.25 was introduced quickly adopted the new protocol. America's Telenet adopted X.25 in 1976, and the Canadian PTT began operating its X.25 network, Datapac, early in 1977. The following year Datapac and Telenet were connected using X.25, demonstrating that the new protocols could be used for internetworking. Other X.25 networks followed in France, Japan, and across Europe. In Europe the PTTs' united support of X.25 created "considerable pressure on computer manufacturers to provide X.25 hardware and software" and made suppliers of private networks more inclined to provide X.25 interfaces.[6] IBM, Digital Equipment Corporation, Honeywell and other American suppliers announced that they, too, had plans to support the new protocol.[7]

The Debate Between X.25 and TCP/IP

Each of these two protocol suites had its own constituency. TCP/IP was popular in the United States, in the international research community, and among users with complex and heterogeneous computing facilities. X.25 was more widely adopted outside the United States and was the choice of public data networks and thus of their customers. Because these constituencies had different aims, assumptions, and priorities, each regarded the technical choices of the other with a certain amount of incomprehension and even hostility.

Nowhere is this more apparent than in the prolonged debate over standards that arose within the data communications community. From 1976 until the late 1980s the merits of X.25 and TCP/IP were debated at computing conferences and in a steady stream of articles in the trade press and in professional journals such as *Computer Networks*, *Proceedings of the Institute of Electrical and Electronic Engineers*, and *Communications of the Association for Computing Machinery*. Reading these arguments can be a disorienting experience, as the same terms (such as *internetworking*) are often used to indicate radically different concepts. Arguments tend to be framed as the evaluation of trade-offs between different technical costs and benefits, yet the assessment of these trade-offs depends on tacit assumptions about the goals and operating environment of the networks. A few examples will illustrate how technical arguments were often the means of expressing social or organizational preferences and concerns.

How Uniform Can Networks Be?

If networks are to function together as an internet they must have some common procedures or protocols to handle addressing, routing, and end-to-end reliability. The PTTs' developers wanted this common element to be the network protocol, which specifies the service provided by the subnet; they felt that connecting networks would be easiest if the internal operations of each subnet were the same. But TCP/IP advocates believed that standardizing at the level of the subnet was unrealistic. Different techniques had been optimized for different media, they argued, and no single network protocol could be best for all. Moreover, the particular type of network protocol described by X.25—a virtual circuit network—would require reliability and processing power in the subnet, which some networks could not supply.

The TCP/IP approach was to have the common element be a separate internet protocol. The Internet Protocol (IP) specified only those functions specifically required for internetworking (i.e., packet and address formats), thus minimizing the degree of uniformity required. But X.25 advocates argued that IP was an unnecessary complication: if all participating networks used the same network protocol, there would be no need for a separate internet protocol. CCITT did add an internet protocol, called X.75, in 1978, but this was only a slight variation of X.25 and assumed that each network was running X.25 or a similar protocol.[8] Also, X.75 was only allowed for connections between public networks, so private networks still had to interconnect using X.25.

The choice between having a uniform network or internet protocol represents a trade-off between the simplicity and efficiency of having uniform networks and the flexibility of allowing diverse ones. The PTTs chose a higher degree of uniformity because they assumed that their monopoly on telecommunications would be extended into public data networking. With most publicly-accessible networks under PTT control, it would be feasible for them to agree on and enforce a single network standard. They also reasoned that having a standard network protocol would be in the public interest: manufacturers could build their products to a single set of specifications, and customers could use the same equipment to connect to any public network.[9] In contrast, attempts to connect dissimilar networks were viewed as a "temporary stopgap" on the way to a desired uniformity.[10]

U.S. network builders, on the other hand, assumed that an internet would have to accommodate different types of systems. Commercial and government-funded networks using various technologies were being built across the country; rapid deregulation in the United States was opening up communications monopolies, and several different public data networks had been launched by the mid-1970s. In the face of this diversity, it seemed impossible to convince everyone to use the same network protocol. It was easy to argue that the more practical course was to rely on a standard internet protocol, such as IP, that could run on top of any underlying subnet. In addition, some network researchers, including those at ARPA, felt that different subnet technologies were necessary to allow networks to be tuned to particular performance requirements. In their view, interfacing heterogeneous networks was not just a stopgap measure until networks could be made uniform, but a permanent necessity.[11]

Ensuring Network Reliability

Another dispute was over the relative reliability of the two systems. The PTTs saw their system as clearly more reliable, since the X.25 network protocol guaranteed error-free delivery of packets from end to end. Since CCITT was confident that public data networks could provide consistently high-quality service, it saw additional reliability measures as a needless complication.[12]

ARPA, on the other hand, was prepared to include in its internet networks that were known to be less reliable but that offered offsetting advantages such as simplicity, low cost, or mobility. ARPA also wanted the capability to use data communications under the hostile conditions of war, when even a normally reliable network might suffer damage to some of its underlying hardware. Therefore, ARPA regarded a network as reliable only if it could provide high-quality service independent of the quality of the underlying subnet. TCP was designed to make hosts responsible for correcting errors no matter what the supposed reliability of the underlying network, allowing the system to tolerate some data loss within the subnet. In fact, since TCP users believed that hosts had to be able to compensate for potentially unreliable networks, they felt it was a wasteful redundancy to have elaborate error control procedures in the network protocol, as X.25 did.[13]

Essentially the debate over reliability hinged on assumptions as to whether future networks would have uniformly high-quality subnets or whether some would provide a lesser degree of service. The arguments for X.25 and TCP/IP both make technical sense given a particular world view, but neither side spent much time discussing the assumptions upon which they based their claims.

The Role of Host Computers

The alternative to relying on the subnet for end-to-end flow control, error control, and virtual circuits is to rely on the hosts. The ARPA protocols had the hosts run a separate transport protocol, TCP, to perform these functions; X.25 required no such protocol. Arguments over the need for a transport protocol stemmed from conflicting expectations about the demand for network services and preferences concerning the role of host computers.

The PTTs saw computer communications as similar to telecommunications. In the telephone business, priority had long been placed on providing basic universal service for ordinary users. By analogy, the PTTs expected that typical computer users would communicate through a public network, rather than building their own private systems, and that they would want minimal responsibility for network services. They assumed that networks would be used to connect terminals to computers, not to communicate between computers or between entire local networks.

In contrast, ARPA came from the perspective of military research and development, which tended to favor high-performance systems with relatively sophisticated operators, and the agency was accustomed to operating its own networks. ARPA expected that users would be willing to take on additional complexity in return for greater control; therefore, they favored greater responsibility for host computers. ARPA also argued that, for some applications, the guaranteed reliability of a virtual circuit network would exact a performance penalty. Setting up a circuit delays the transfer of data, as does correcting transmission errors. In real-time applications such as transmission of voice or video, it would be better to accept an occasional error than to introduce a delay in the flow of information. If hosts had a greater role in determining the level of network service, users could choose whether to maximize speed or accuracy.

The PTTs were adamantly against dividing control over network operations between the subnet and the hosts. They felt that while this might be feasible for a private network in which one organization was in charge of both hosts and subnet, divided control would make it difficult for a public network to make service guarantees to its customers. For instance, X.25 provided the ability to regulate data flow on each virtual circuit. TCP had a less sensitive flow control mechanism, and the hosts were responsible for exercising it. X.25 supporters questioned whether hundreds or thousands of host operators running many different versions of TCP could all be counted on to implement this mechanism correctly.[14]

The PTTs also argued that X.25 made networking easier for users, especially those who had not already invested in networking software, because it made minimal demands on their hosts. In their view, small users would be better served by a system that gave them efficient network service without requiring a major upheaval of their systems. Even Lawrence Roberts, who had headed the original ARPANET project, felt that it was impractical to require transport protocols for public networks, and recommended using X.25 instead.[15] The carriers agreed that some services for private groups were needed, but they were in favor of services controlled by themselves. For instance, X.25 provided a "facility field" in its packets that could be used by the hosts to carry control information, but before actually using this option the host would need special permission from its carrier.[16]

In the United States, however, owners of large computer installations in government, universities, and the private sector were used to having complete control over their networks, and they expected to retain an active role in running the system. Large users in many countries wanted the carriers to provide cheap, basic transmission services on which they could build their own systems; some suspected that the PTTs insisted on providing virtual circuit service because they could charge more for it.[17] For sophisticated users, the need to run transport protocols was not an insurmountable obstacle, especially since implementations of TCP/IP were becoming widely available.

The Role of Private Networks

A final source of contention was the charge that X.25 made inadequate provision for the myriad of private networks being built by universities, corporations and government agencies. One problem was that X.25 was difficult to adapt to the technical characteristics of local-area networks (LANs). Local-area network systems such as Xerox's Ethernet and IBM's token ring system were developed in the late 1970s and early 1980s and offered a relatively easy and affordable option for connecting computers at a single location. Instead of having switching nodes connected by transmission links, these LANs broadcast messages across a medium such as a cable that is shared by all the hosts.

One of the arguments made in favor of the X.25 network protocol was that it removed complexity from the hosts by putting most of the functionality in the switching nodes. The philosophy of broadcast LANs, however, is just the opposite. Broadcasting reduces the cost and complexity of LAN technology by eliminating the subnet nodes, leaving hosts with the task of sorting out connections. Since it has no nodes to maintain virtual circuits, it is not practical for a LAN to use X.25 internally, or to provide the connecting link between two X.25 networks.[18] LANs that need to connect to an X.25 network generally designate a single host to translate between the LAN protocol and X.25. In contrast, since the ARPA protocols rely on hosts to create virtual circuits, it is no more difficult to implement TCP/IP for a host on a broadcast network than for any other host.

Another issue was addressing. Hosts on an internet need a way to translate between their local address format (e.g. Ethernet addresses) and the address format of the internet. From an early date, implementations of IP had included software to translate between IP and various address formats commonly used in local-area networks, but standards for doing this for X.25 hosts were not written until much later. Another problem was that the CCITT addressing system, called X.121, allocated far too few network identification numbers for the number of private networks being developed (for instance, only 200 network numbers were allotted for the entire United States).[19] The addressing issue, contends the author of one widely-used networking textbook, "is not one of poor estimation; it is a question of mentality. In CCITT's view, each country *ought* to have just one or two public networks. . . . All the private networks do not count for very much in CCITT's vision."[20]

At the heart of the arguments over private networks are two contrasting assumptions about the environment in which large networks will operate. The X.25 designers expected each host to be directly connected to an X.25 network. They did not anticipate that many users would be attached to multiple networks, so they did not design their protocols to work with alternative systems such as LANs. The carriers envisioned themselves as the only networks; their model of internetworking was to have one large network in each country, connected at national borders to other X.25 networks. In contrast, the ARPA model of internetworking was a set of peer networks connected in a complex web, in which a host might need to transfer data across a whole series of diverse systems. Therefore, they anticipated the need to accommodate different network technologies and addressing systems.

Beyond Debate:
Adapting to a Changing Network Environment

One lesson to be learned from the standards debate is that network protocols have no single, universal meaning. They can only be evaluated in terms of particular users' needs, and there may be as many different evaluations as there are types of users. Thus there was no simple resolution to the internet conflict: rather, there were a number of different responses, depending on how network builders perceived the changing demands of the networking environment.

X.25: *Broadening the Offerings*

As one response, CCITT made a series of revisions to X.25, with new versions appearing in 1980, 1984 and 1988. Even within CCITT there had been members who were in favor of datagram networks and who thought that in the rush to get a standard developed, the X.25 committee had chosen a virtual circuit system without allowing time to consider other options.[21] These members succeeded in adding an optional datagram mode to the X.25 specification, but—unsurprisingly—it was not actually implemented by any public carrier and was removed from the standard in 1984. Instead, a compromise capability called "fast select" was instituted that allows users the option of faster connection set-up at slightly lower reliability. In making this compromise CCITT was acknowledging that users might want to make different trade-offs between speed and reliability.

Other changes allow (though they do not require) X.25 networks to offer more control to users. These include allowing the user to specify the sequence of networks that a connection should go through; allowing the user to send a greater number of packets at one time, so as to make more efficient use of high-speed links such as satellites; and various facilities to aid in connecting private networks, such as more flexible addressing and private diagnostic codes. There is also a "closed user group" option that allows subscribers to specify that a set of hosts will only accept messages from each other, as if they resided on a separate private network.[22] CCITT saw these services as combining for users the advantages of having their own network with the efficiency of sharing public facilities. They also represent a strategy for preserving the PTTs' ultimate control over the network system.

The public carriers have also proposed new alternatives to X.25 that are intended to better meet the needs of computer users, especially owners of local-area networks. Techniques such as "frame relay" and "asynchronous transfer mode (ATM) switching" were introduced during the 1980s as a way to provide high-speed connections between local-area networks without putting undue technical burdens on them. Adopting new technologies allows the PTTs to respond to shifts in the type of service demanded (e.g. from terminals to local-area networks) while keeping their own networks at the center of the internet system.

TCP/IP: Strategies for Connection

While the PTTs were trying to incorporate different user services within their own network systems, ARPA was exploring ways to exchange data between different types of networks. In the ARPA internet scheme, networks are connected by *gateways*, computers that are attached to two or more networks and act as the interface between them. The common language of these gateways is the Internet Protocol (IP); IP gateways determine how to route packets between networks and can be used to connect any combination of local, regional, and wide-area networks. But gateways can also be used to translate between different network protocols. Translation gateways thus fit David and Bunn's description of a "gateway technology," defined as

some means (a device, or a convention) for effectuating whatever technical connections between distinct production subsystems are required in order for them to be utilized in conjunction .... A gateway technology, therefore, achieves technical compatibility in order to affect linkage or communication among subsystems.[23]

A computer that ran both TCP/IP and X.25 could be placed between two networks as a gateway, providing one alternative for connecting the two types of networks. However, such arrangements are inefficient because of the amount of processing the gateway has to perform to mediate between two dissimilar network protocols.

Another approach is to have the host computers on an X.25 network run TCP/IP over the X.25 subnet. In 1982 ARPA experimented with this technique by having TCP/IP hosts communicate with each other over the Telenet X.25 network.[24] While it may seem like a simple technical fix, this approach is radical in its implications, because it reconstructs the rival X.25 system as a subsidiary component of the TCP/IP system. CCITT, it should be recalled, designed X.25 to be the primary provider of end-to-end network control. By running TCP/IP *over* X.25 and relying on TCP for end-to-end control, ARPA reduced the role of X.25 to providing a simple data conduit. TCP/IP users no longer even see TCP/IP and X.25 as alternatives, arguing that they function on different levels of the network system.[25] They thus define the standards conflict out of existence by ignoring X.25's claim to be a complete system for networking and internetworking.

A third strategy for interconnecting incompatible systems has been extremely influential in shaping the perception and use of internetworking. This approach ignores differences at the subnet level and instead focuses on translating between higher-level protocols, particularly electronic mail systems. By 1982 ARPA was experimenting with translation gateways that would link incompatible mail systems by decoding and re-encoding mail messages that passed between networks.[26] Other research and commercial organizations have also been active in devising mail gateways, making it possible to exchange messages between all of the most widely-used mail systems.

The result has been that electronic mail is the only form of connection that has been fully achieved between diverse networks.[27] While the potential demand for other services, such as file transfer or remote login, may have been just as high, mail was much simpler to implement since it does not require an interactive connection between host computers. Thus the communications aspect of networking has been emphasized, encouraging new uses of computers that stress interaction rather than calculation. Mail has shaped the perception of computer and network technologies among non-expert users, whose sole use of the computer may be for services such as electronic mail and bulletin boards. Internet mail has also increased the size and scope of the communities that can be formed using electronic communications. The technology itself does not dictate these outcomes, but it has made some options more attractive and indeed more imaginable than others.

Toward a Common Framework: Open Systems Interconnection

No account of the evolution of network standards would be complete without mentioning the efforts of the International Standards Organization (ISO). Noting the emergence of competing and incompatible "standards" for networking, ISO decided in 1977 to try to rationalize the networking debate by laying down a framework within which protocols could be developed. The project was called Open Systems Interconnection (OSI), because ISO believed that making network specifications "open," or independent of any proprietary computer system, would prevent users from being locked into a single manufacturer's product line.

The OSI project was not meant to develop actual protocols like X.25 or TCP/IP, but to provide an abstract model that would define what services protocols should provide and how they should interact. ISO described the project as a way to "position existing standards in perspective," "identify areas where standards should be developed," and "expand without disrupting previously defined protocols and interfaces."[28] Once ISO had constructed the model, which described seven layers of network protocols, it would begin a slow process of examining existing or proposed protocols and selecting one or more as the approved standard(s) for each layer.

ISO members voted to make OSI an International Standard in 1983, and it became the dominant model for subsequent discussions of networking. Many national governments, as well as the U.S. Department of Defense, resolved to switch their networks to OSI-conformant protocols. Network builders and users also tried to win ISO approval for their protocols: eventually both X.25 and IP, as well as the major local-area network protocols, were incorporated into the OSI scheme.

While the Open System Interconnection model provides a common framework for discussing protocols, it has so far failed to resolve conflicting approaches to networking. One reason is that even networks that use OSI protocols tend to implement only a subset of the total OSI system. Thus many X.25 networks have chosen not to implement the OSI transport protocols, just as they declined to use the TCP transport protocol. Another reason is that the International Standards Organization has been reluctant to reject protocols that have already become *de facto* standards; when two or more established protocols have been proposed for the same layer in the OSI model, ISO has often adopted all of them. This means that it is possible for two networks to use OSI-conformant protocols and still be incompatible. Finally, despite official government pronouncements, network owners have been in no hurry to replace their existing systems, especially since concrete protocol specifications for many of the OSI layers have been slow in coming. OSI may be more significant as a symbol of the widespread desire for standards than as an alternative to existing protocol systems.

Conclusion

Two decades since international standards were first proposed, there still exist a host of widely-used "standards." These include manufacturers' products like DECNET or IBM's System Network Architecture, local-area network systems like Ethernet and token ring, X.25 and other carrier standards, TCP/IP, and a number of others. At the same time, networks with little else in common are able to cooperatively provide the basic service of electronic mail to users around the world. Whether a single protocol system, such as that proposed by ISO, will ever be universally adopted, or whether gateway services like mail will continue to bridge the gap between different networks is not clear. It seems likely, however, that there will be a diversity of networking technologies as long as there are significant differences in the communications needs of computer users. And while future controversies in computer networking may center around different issues, such as privacy, access, or cost, standards will continue to be contested for their social implications as well as their technical merits.

When large technological systems undergo change, the process often involves diverse actors, increasing the likelihood of conflict and incomprehension between different groups. A knowledge of the practices, assumptions and goals of these different actors can help uncover what is really at stake when seemingly technical criteria such as "complexity" or "reliability" are debated. Reconstructing the cultural reference points of system builders can help the historian to identify where the gulfs lie between competing technologies and to understand the strategies of those who attempt to cross them.

List of Abbreviations

CCITT   Comité Consultatif International Télégraphique et Téléphonique (Consultative Committee on International Telegraphy and Telephony)

ARPA	Advanced Research Projects Agency (U.S.)
PTT	Post, Telegraph and Telephone Administration
TCP/IP	Transmission Control Protocol/Internet Protocol
ISO	International Standards Organization

Notes

1. See e.g. David and Bunn, 1988.

2. Hirsch, 1976a; Hirsch, 1975.

3. Since the United States does not have a national carrier, it is represented in CCITT by the State Department.

4. Cerf, 1992.

5. Rybczynski et al., 1976; Dhas and Konangi, 1986.

6. Davies and Bates, 1982, p. 20.

7. Hirsch, 1976a.

8. Cerf and Kirstein, 1978, p. 1402.

9. Rybczynski et al., 1976, pp. 7, 8; Danet et al., 1976, p. 253.

10. Quarterman, 1990, p. 191.

11. See e.g. Cerf, 1990, p. 38; Cerf and Kirstein, 1978; Callon, 1983.

12. Quarterman, 1990, p. 191.

13. Cerf and Kirstein, 1978, p. 1403.

14. Danet et al., 1976, p. 251-2; Roberts, 1978, p. 1,310.

15. Roberts, 1978, p. 1,310.

16. Hirsch, 1976b.

17. Tanenbaum, 1989, p. 322; Hirsch, 1976a; Pouzin, 1975.

18. Tanenbaum, 1989, p. 346.

19. Cerf and Kirstein, 1978, p. 1,400.

20. Tanenbaum, 1989, p. 322.

21. Pouzin, 1975.

22. Danet et al., 1976, p. 252.

23. David and Bunn, 1988, p. 170.

24. Postel et al., 1982, p. 978.

25. Cerf, 1992.

26. Postel et al., 1982, p. 978.

27. Quarterman, 1990, p. 191.

28. ISO/TC 97/SC 16, 1978, p. 50.

References

Callon, Ross. 1983. "Internetwork Protocol." *Proceedings of the IEEE* 71(12): 1388-1393.

Cerf, Vinton. 1990. Interview by J. O'Neill for the Charles Babbage Institute, April 24. Reston, VA.

_____. 1992. Personal communication, January 17.

Cerf, Vinton G., and Peter T. Kirstein. 1978. "Issues in Packet-Network Interconnection." *Proceedings of the IEEE* 66(11):1386-1408.

Danet, A., R. Despres, A. LeRest, G. Pichon, and S. Ritzenthaler. 1976. "The French Public Packet Switching Service: The Transpac Network." In *3rd International Conference on Computer Communication*, ed. P. K. Verma, 251-259.

David, Paul A., and Julie Ann Bunn. 1988. "The Economics of Gateway Technologies and Network Evolution: Lessons from Electricity Supply History." *Information Economics and Policy* 3:165-202.

Davies, B. H., and A. S. Bates. 1982. "Internetworking in the Military Environment." *IEEE Computing Society International Conference (Compcon)*. 19-29.

Dhas, C. R., and V. K. Konangi. 1986. "X.25: An Interface to Public Packet Networks." *IEEE Communications Magazine* 24(9):18-25.

Hirsch, Phil. 1975. "Canada Network Won't Take SDLC Protocol." *Datamation*. (March) 121-123.

———. 1976a. "Protocol Control: Carriers or Users?" *Datamation*, March: 188-189.

———. 1976b. "Protocol for Packet Networks: The Question is Implementation." *Datamation* (May) 187-190.

ISO/TC 97/SC 16. 1978. "Provisional Model of Open-Systems Architecture." *Computer Communication Review (ACM)* 8(3):49-62.

Postel, J., C. Sunshine, and D. Cohen. 1982. "Recent Developments in the DARPA Internet Program." In *Pathways to the Information Society*, ed. M. B. Williams, 975-980. New York: North-Holland Publishing Co.

Pouzin, Louis. 1975. "The Communications Network Snarl." *Datamation* (December) 70-72.

Quarterman, John S. 1990. *The Matrix: Computer Networks and Conferencing Systems Worldwide*. Burlington, Mass: Digital Press.

Roberts, Lawrence. 1978. "The Evolution of Packet Switching." *Proceedings of the IEEE* 66(11):1307-1313.

Rybczynski, A., B. Wessler, R. Despres, and J. Wedlake. 1976. "A New Communication Protocol for Accessing Data Networks: The International Packet-Mode Interface." *International Conference on Communications*. 20, 7-11.

Tanenbaum, Andrew S. 1989. *Computer Networks*. 2nd ed. Englewood Cliffs, NJ: Prentice Hall.

10

Integrating Supple Technologies into Utility Power Systems: Possibilities for Reconfiguration

Alexandra von Meier

Introduction

Solar energy conversion and related technologies for dispersed generation and control of electric power are emerging as promising options to support and augment existing power systems. These "supple" technologies in many ways resemble "soft" energy technologies, which have been thought incompatible with the traditional, centralized approach to power systems. In light of recent developments in the energy and electric utility industry, however, supple technologies may offer constructive and elegant solutions to problems of established power systems rather than necessarily competing as alternatives.

This chapter addresses the possibility of an integration of the "hard" and "soft" approaches, encompassing the evolution of the physical system with its hardware and interconnections, planning and evaluation methods, and the institutional and cultural context in which they are employed. Such an integration may constitute a radical reconfiguration of existing power systems — not a rapid transformation resulting from a conflict or struggle between two opposing strategies but rather a gentle subversion or an adiabatic change resulting from steps taken to minimize costs and system strain.

Terminology

The terminology of "hard" and "soft" technologies and energy supply strategies was introduced by Amory Lovins. In his analysis and critique, a "hard" energy system was characterized by large fossil-fuel or nuclear power plants and a centralized, hierarchical structure of the physical transmission and distribution system and the institutions that manage them. Society's commitment to growing energy consumption and the concentration on a few powerful but ultimately vulnerable resources and technologies was termed the "hard path." In contrast,

> Soft technologies are defined by five characteristics:
> 1. They rely on renewable energy flows.
> 2. They are diverse so that...national energy supply is an aggregate of very many individually modest contributions, each designed for maximum effectiveness in particular circumstances.
> 3. They are flexible and relatively low technology — which does not mean unsophisticated but rather easy to understand and use without esoteric skills, accessible rather than arcane.
> 4. They are matched in *scale* and in geographic distribution to end use needs, taking advantage of the free distribution of most natural energy flows.
> 5. They are matched in *energy quality* to end-use needs.[1]

Lovins characterized this as "a textural description, intended to mean...flexible, resilient, sustainable and benign."

"Soft energy paths" were defined as those dependent on soft technologies. This definition involved the anticipated consequences of putting technologies with these characteristics into use since "the social structure is significantly shaped by the rapid deployment of soft technologies." Specifically, "[t]he distinction between the hard and soft energy paths rests not on how much energy is used but rather on the technical and political *structure* of the energy system, thus focusing our attention on consequent and crucial political differences."[2]

The technologies termed "supple" in this chapter overlap to a large extent with Lovins' soft technologies, both with respect to their descriptive characteristics and the specific examples that would be included to illustrate the genre (such as photovoltaic cells or wind turbines). Supple technologies generally share the above soft characteristics, though not necessarily in every case. For example, some technologies that can be considered supple (such as fuel cells or small-scalle gas turbines) still rely on fossil fuels rather than on renewable energy flows. Thus, rather than stretching Lovins' definition to suit the present purpose, the term "supple technologies" is proposed as an alternative.

Supple technologies are better described by the following physical characteristics:

1. They are modular, meaning that they can be constructed, operated and maintained effectively on a range of scales. They defy an engineering definition of "optimal scale" based on their physical properties or the logistics of their deployment and operation. Their optimal scale is instead context dependent.
2. As a result of modularity, they can be sited in a dispersed fashion. Dispersal can be geographic and tied to the functional position in the electric grid.
3. They are flexible. They can be sited, constructed and brought online within a short period of time and with a low risk of failure.

These physical characteristics are the basis for the defining property of supple technologies: they do not require or favor a centralized and hierarchical structure of the system in which they are embedded, in either a technical or institutional sense. Because of their modularity and the possibility of geographic dispersal, supple generation technologies do not give rise to a natural monopoly in electricity production. Dispersed generating facilities may be owned and operated by a single utility company or by a large number of smaller independent actors.

In addition to modular generating technologies, dispersed generation within the grid requires a transmission and distribution (T&D) system capable of handling or even taking advantage of the inflow of power from dispersed and heterogeneous sources. T&D devices that facilitate dispersed operation are therefore included in the definition of supple technologies. These devices include T&D hardware (switches, sensors, relays), communication equipment, and algorithms for dispatching generation and reconfiguring circuits.

Given the technical possibility of dispersed coordination, dispatch and control, the ensemble of supple technologies does not support a natural monopoly in the transmission and distribution of electricity. Since it is based on a capital-intensive infrastructure, the transport of electricity has been considered a natural monopoly even if the ownership and operation of generation is decentralized. However, "smart" T&D devices make it theoretically possible to operate a power system without strong centralized control and coordination. It is therefore conceivable that even the natural monopoly of the distribution of electricity could erode. The defining characteristic of supple technologies — that they do not require or favor a centralized system structure — thus applies to generation as well as to the T&D system, in both technical and institutional terms.

There are three differences between the term "supple" in this usage and Lovins' definition of "soft." First, a minor difference lies in the breadth of the definition. Although there is significant overlap for generation technologies, "supple" is applied more broadly to include T&D equipment and energy storage devices. Conservation and load management can also be considered supple options on a menu of planning choices. Second, while "soft" emphasizes reliance on renewable resources and focuses on energy flows through the system, "supple" emphasizes modularity and flexibility, focusing on the morphology of the system itself. The third and most important difference is one of interpretation and has to do with the assumed consequences of implementing these technologies. By calling a technology supple, it is not implied that its implementation will *cause* a change in the institutional, social and political context of the energy system. Based on its physical and operational characteristics and the logistics of its deployment, however, a supple technology *is consistent with* such change. It *does not preclude or discourage* change. In other words, supple technologies are indifferent to certain aspects of the larger system in which they are embedded.

New Technologies

Electric utilities today can employ a number of supple technologies as a result of recent technological advances, cost reductions and operating experience.

Photovoltaics (PV) may be considered an archetypal supple technology: the size of existing PV systems ranges over eight orders of magnitude. They impose minimal constraints on a site and its surroundings. Photovoltaics are becoming a realistic power generation option for utility-connected applications because their technical feasibility and practicality has been established in niche markets and many large-scale demonstration projects, and their cost continues to decline. Important technological development has occurred in the area of inverters, which are the interface between a direct current power source such as a PV array and an a.c. grid. Until recently, many utility engineers felt that inverters could not be operated safely and successfully in distribution systems. Due to dramatic improvements in inverter performance and the accumulation of utility operating experience, these views are changing. New inverters are expected to actually enhance power quality in distribution systems.

Additional progress has been made with other supple generation technologies and their integration into the grid. Wind turbines have seen significant cost and performance improvements over the past two decades. Solar thermal electric generation has also proven to work reliably and is on the verge of profitability in the current economic environment.

Parallel to the development of renewable options is a move toward smaller unit size in fossil-fuel based power generation, marking the end of an almost century-long trend to exploit economies of scale by building ever larger turbines and power plants. Gas turbines, for example, are becoming increasingly popular because of their smaller scale, short planning and construction lead times, and responsive operation. Fuel cells offer higher energy conversion efficiency and the greatest degree of modularity among fossil-fuel based generation technologies. They are also approaching commercial feasibility as a result of substantial research and development efforts by utilities.

On the transmission and distribution side, smaller and "smarter" control technologies are being developed and tested that can facilitate the integration of dispersed and heterogeneous generation into the existing grid and enhance the operational and economic benefits of such generation to the system. These distribution automation technologies include solid-state sensors, switches, relays and communication devices. They can refine the control and coordination of dispersed generation and loads in distribution systems, allowing control operations to be performed remotely and much more frequently. They also permit reconfiguration of distribution circuits, changing the system topology in real time according to varying load conditions. A variety of distribution automation technologies have been or are in the process of being technically proven and are now becoming available. Though their use by U.S. utilities is still minimal due to high costs, many utility planners expect to use such technologies more often over the coming decade.

Problems and Constraints of the Hard System

Electric power systems in the U.S. are currently facing technical and institutional problems and constraints that may motivate change. These problems arise both from the systems' physical or technical characteristics and the economic and regulatory environment in which the physical systems operate. The resulting financial constraints may curb utilities' willingness to experiment with new technologies and planning methods. But because these problems signal inadequacies of the traditional "way of doing business," they may simultaneously exert pressure on the industry to reform.

Planning and installing new generation capacity has become increasingly difficult for utilities since the late 1960s. Capital costs for new plants have increased for two reasons. Design improvements, efficiency gains and economies of scale in the development of boilers and large steam turbines no longer allowed for declining costs as they had throughout the history of the industry.[3] Also, heightened environmental awareness has increased the complexity of siting and obtaining power plant licenses, which in turn has required more elaborate hardware, legal costs, and greater financing costs during planning and construction. Environmental concerns specifically have forced utilities to invest in strategies to mitigate emissions from fossil-fuel combustion and to pursue lower-impact technologies based on renewable energy.

The longer lead times of generation projects due to the above complications also compound the challenge of electric supply planning given uncertain demand forecasts. New power plants relying on traditional technologies have become less flexible as investment projects. In the 1970s, the growth rate of electric demand declined dramatically and since then has remained bounded but difficult to predict. Finally, the volatility of fossil-fuel prices translated into higher fuel costs and less certainty in economic planning. The combination of these circumstances led many U.S. utilities into financial crises.

Transmission and distribution (T&D) capacity constraints have also gained importance recently. Until the 1980s, the foremost concern among utility planners was to assure that sufficient power generating capacity would be available to meet demand, while transmission was only a minor issue. Indeed, during the decades of expansion, utilities generously oversized their transmission systems. They were overwhelmed by the formidable analytic challenge of rigorously and quantitatively relating transmission capacity to system reliability and they could afford to err on the safe side. Thus, the classical methods of system reliability evaluation assume that the transmission system is capable of carrying power flows from generation sources to load points within an area whenever needed.[4] More and more often, though, the ability to provide power to customers depends not on the ability to generate or procure power but on whether this power can be delivered to the desired location.

Transmission and distribution capacity is becoming as much — if not more — of a limiting factor than generation as a result of several interconnected circumstances. First, utilities' current financial constraints prohibit the generous redundancies of the "grow-and-build" era. There is now economic pressure to utilize existing assets as efficiently as possible and avoid new construction until proven necessary. Second, while most utilities have provided for sufficient generation capacity reserve margins on a systemwide basis in anticipation of load growth, load growth often takes place in specific localized regions. Third, T&D upgrade investments are unattractive to utility planners, shareholders, regulators and ratepayers because they involve high capital costs without promising new revenues or an easily measurable improvement in the quality of service. At the same time, the construction of new transmission lines may be opposed on environmental or aesthetic grounds, making it more difficult for utilities to obtain rights-of-way.

Finally, greater demands are placed on existing T&D capacity because power systems are widening their scope for supply sources and are increasingly relying upon purchased power that is imported over greater distances. Even utilities who do not themselves purchase power are affected by this development. They can offer their transmission capacity to neighboring utilities in exchange for a transportation or "wheeling" charge. Wheeling contracts, which have become increasingly prevalent and subject to heated negotiations since the 1980s, signify the value of transmission capacity *per se* and its scarcity in the larger network of systems.

How Supple Technologies Can Help

Supple technologies can help alleviate these problems and constraints within existing power systems when they are implemented with an integrative planning perspective. They can support the system fall by allowing temporal and geographic planning flexibility, relieving transmission and distribution constraints, and reducing electric losses.

Planning flexibility is the result of modularity and short lead times for generation technologies. Flexibility in this context means that the system can be altered in a short time or on a limited regional basis in response to different circumstances such as changes in electric demand or price and availability of energy resources. A flexible system reduces the dependence on long-term demand forecasts because new capacity (generation as well as T&D) can be brought on-line quickly in response to needs. Planning flexibility protects utilities from both poor asset utilization due to oversized capacity and falling short of supply.

Another planning advantage of supple generation technologies based on renewable energy is reduced financial uncertainty. Though their capital costs tend to be high, solar and wind power plants offer a high degree of certainty regarding their lifetime cost. Unlike traditional fossil-fuel or nuclear plants, solar and wind plants have no fuel cost, low probabilities of licensing and construction delays, and low vulnerability to failure of individual components. Fewer things can happen to a well-designed solar or wind power plant that will threaten availability or the construction and operation budget.

Although they do involve uncertain fuel cost, modular fossil-fuel technologies still have the advantage of smaller size and accordingly a spreading of risk.

Supple technologies can also relieve transmission and distribution constraints. For example, distribution automation technologies would allow the reconfiguring of circuits to shift load away from overloaded components, providing the overload is localized, which is often the case. An even more potent tool for alleviating T&D constraints is the strategic placement of generation in the distribution system or at substations. Feeding power into the grid at a location near the customers reduces the current flow from central-station power plants or backbone transmission lines to the location of the dispersed generation. Thus, if the dispersed generation is connected on the customer side or "downstream" of a transmission or distribution bottleneck such as a transformer or a distribution line, it relieves the constraint of power flow through this bottleneck. The actual generation site could be at a transmission or distribution substation or in the case of photovoltaics, even on customer rooftops. It is crucial, of course, that the generation capacity is available during the peak demand hours at which the T&D constraint would be felt. For utilities with day- and summer-peaking loads, the solar resource is ideally suited.[5]

Finally, supple technologies can help decrease energy losses by reducing the electric current flow on heavily loaded lines. Distribution automation can accomplish this by reconfiguring circuits in real-time to balance their loading. Dispersed generation reduces losses in both the transmission and distribution systems simply by providing power closer to the customers, so that there is less current traveling long distances. The reduction of losses represents a direct energy savings for the utility and helps avoid capital costs of capacitors and other equipment used to balance voltage levels in distribution systems. Though line losses and voltage control are not considered an urgent problem in the industry since they are predictable expenses that constitute a relatively modest fraction of utility budgets,[6] they represent an area in which supple technologies can provide clearly measurable economic benefits.

Reconfiguration by Integrating
Supple Technologies

The growing technical and economic feasibility of using supple technologies in electric power systems opens up a wide range of possibilities for the reconfiguration of these systems. There are five distinct realms of operation in which such reconfiguration can take place. The first realm embraces the physical system, including the hardware for generation, transmission, storage and end-use, as well as the energy flowing through it. The second realm covers the control of the system: the dispatching of power, switching of circuits, the communication links and the algorithms used for control. The third realm involves system planning: the framework for decision-making, how objectives are defined, what options are compared to what, and by which criteria. The fourth realm is that of ownership, law, and regulation: who owns equipment, who makes operating and planning decisions, and the nature of the relationship among the different actors. The fifth realm is that of management and culture: how individuals work together and how they perceive and understand the system they are working on.

The major implication of supple technologies in the physical realm is that generation can be smaller and more dispersed, connected to the grid at various voltages and sited specifically to alleviate T&D constraints. Intermittent generating resources such as solar and wind could be complemented by storage on location for output leveling. Power flow on transmission lines would be reduced. More generally, the strength of the connections between regional pieces of the network might vary; some areas could even be intermittently or permanently operated as "power islands" isolated from the remainder of the grid. This physical decentralization of the grid is made possible by modular generation in conjunction with distribution automation technologies that facilitate their dispatching, their integration with electric demand, the switching of circuits, and the coordination of electrical protection (e.g. circuit breakers).

In the realm of grid operation and control, supple technologies offer an increased degree of refinement compared to current grid management strategies. They monitor and coordinate the interaction of more heterogeneous components. The combination of dispersed generation and distribution automation opens up the possibility of dispersed dispatching on the basis of local measurements, in addition to or as an alternative to central dispatching relayed through telecommunication. Distribution automation technologies can also provide more refined and interactive connections with customers, such as time-of-use metering and even the control of certain loads by the utility for the purpose of leveling demand.

In the planning realm, a variety of changes are conceivable but their relationships with supple technologies are not entirely clear. The short lead time of supple generation technologies allows a shorter time span for generation planning and thus a more flexible response to load changes. A logical extension of this planning flexibility would be the integration and equal consideration of supply and demand-side options, which may be aimed toward conservation or shifting load to off-peak hours. Another aspect of more integrated planning is the joint consideration of generation and T&D, which is essential for the evaluation of dispersed generation because costs and benefits may apply to different areas. These kinds of integration imply organizational reform within utilities, which currently tend to plan and analyze generation, T&D, and demand-side management projects in separate departments.

It should be noted here that flexible and integrated planning involves much detailed information about the system, just as the dispatching of many small generating units and loads and the reconfiguration of distribution circuits assume an increased volume and refinement of available information about system parameters. Indeed, the evolution of control and planning of power systems towards greater refinement and localization depends on a shift in the cost of information versus energy. In the past, energy has been cheap and equipment for acquiring specific localized information about the system has been expensive. As the cost of information decreases relative to the cost of energy, there is a greater incentive for sophistication and refinement in the operation and planning of power systems. Reducing the cost of information is the most significant result of advances in supple T&D technologies.

In the legal realm, there is the possibility of dispersed ownership of various components of power systems. The traditional monopoly of electric utilities on building and operating generation capacity has already been broken up in the U.S. by the 1978 Public Utilities Regulatory Policies Act (PURPA) that required utilities to purchase power from non-utility producers at "avoided cost." The resulting contribution to electric capacity by private generation has grown to as much as twenty percent in California in 1990.[7] The role of technology in this development was limited. Though the existence of technologies such as cogeneration that could be operated well and economically by non-utility actors was a necessary condition for the enthusiastic response, it was not sufficient to effect such a significant structural change. Rather, the legal reform itself was crucial. Because as regulated monopolies, utilities are guaranteed a rate of return on their investments, they previously had no incentive to purchase power from private producers even if this was their least-cost option for acquiring capacity or energy.

Supple technologies could play a significant role in the continuing evolution of dispersed ownership of electric grid components. Smaller generation projects would tend to help diversify the constituency of power producers. Cost-effective applications of supple generation technologies for grid support would be recognized. This may also lead to increased overall system penetration of smaller-scale distributed generation projects that lend themselves to non-utility ownership.

The trend toward the vertical disintegration of utilities can be mirrored in horizontal disintegration or regionalization of pieces of the grid, as for example in the recent division and privatization of the formerly national electric power system in the United Kingdom. The U.K. system is now divided vertically into independent generating companies, a national grid company, and a distribution sector, which is in turn divided horizontally into twelve regional "public electricity suppliers."[8] The Western European grid is an example of a synchronous system consisting of a number of diverse interconnected individual subsystems that belong to the different countries and are managed in different ways. Similarly, many individual utilities throughout the Northeastern U.S. are interconnected so as to establish a larger system from regionally divided pieces. These latter two cases do not exemplify disintegration since they have evolved by increasing connection among historically independent systems rather than division and dispersal of a once unified system. They do, however, illustrate the feasibility of horizontally segmented electric systems in a static sense. Though these systems are complex in that they involve many interacting components, their stability and reliability of operation is not compromised by the regional divisions. With regard to supple technologies, an important question is to what extent such networks of interconnected regions can be down-scaled, i.e. how small the independently owned, planned and administered regions within a stable functioning network can be.

Current approaches to opening the electricity market focus mainly on requiring electric utilities to provide open access to their T&D facilities for transactions among private buyers and sellers. It is conceivable, however, that the monopoly of utilities could be broken up completely. For example, independent suppliers could obtain license to construct T&D facilities augmenting the existing grid. It is not obvious what factors would lead to a change of legislation in this direction. The fundamental justification for granting exclusive franchise rights within service territories is unarguably valid, namely that the duplication of transmission and distribution infrastructure would constitute an inefficient use of resources. However, it is also based on the assumption that the tightly coupled components of a T&D system cannot be operated by more than one company without adverse technical implications for stability and reliability. This assumption is no longer necessarily true, given the technologies now available. Thus, supple technologies allow for the possibility of decentralized T&D ownership without the duplication of infrastructure.

The impact of supple technologies is most difficult to estimate in the realm of management and culture. One might expect a decentralized system hardware structure to ultimately lend itself to a similarly decentralized management structure. For example, a large number of generation facilities would need an equal number of organizational units (which may or may not belong to the same company) for their operation and administration. Similarly, the degree of autonomy enjoyed by the management of regional grid sections would depend on the degree to which decentralized operation and planning are compatible with the technology. It is conceivable that following the pattern of the physical system, hierarchical organizational structures would give way to more cooperative and interactive relationships,[9] though the specific mechanisms for transmitting such a change are not immediately obvious.

There are also implications for the relationship between electric service providers and customers or society in general. Specifically, a grid with open access to buyers and sellers ultimately means abandoning the traditional obligation of a utility to serve all customers in its territory. Customers would then be more autonomous actors, less able to depend on the grid and probably more inclined to self-generate with small-scale equipment. The utility would no longer be the universal, trusted service provider. In this case, a change in the physical and organizational system structure also implies a change in the overall system culture.

Given these five realms of electric power system operations and the significance of supple technologies within each, the question arises how strongly they are linked. Can a system undergo a radical reconfiguration within one realm but not others? Or does any significant change in one realm necessarily result in changes in all other realms? Linkages between realms may be indirect and based on the interaction of internal and external factors. Accordingly, the propagation of change across realms may be difficult to predict.

For example, Pacific Gas and Electric (PG&E) has discovered that connecting remote customers to the grid can be more costly than installing self-sufficient, stand-alone photovoltaic systems with battery storage.[10] Though strategies to minimize overall cost would clearly favor PV systems over line extensions, it is not clear from regulations whether an electric utility should be authorized to provide stand-alone power under its mandate to serve. Current U.S. regulation would most likely prohibit unfair competition between the utility with a captive audience of customers and smaller private installers of stand-alone energy systems.

On the other hand, since line extensions up to a certain limit (usually several miles) are currently subsidized by utilities' customer base, customers have no economic incentive to obtain a private PV installation. If economic rationality prevails, stand-alone PV systems will eventually displace extended connections to the grid. There are various possibilities for who would own, install and service these systems (including third parties) and how the economic transfers among the participants would be negotiated. Any of these possibilities would imply a significant departure from the traditional utility-customer relationship. Technically, however, the benefit of such PV installations to the system's overall economic efficiency is independent of legal and institutional arrangements. Thus, while these possibilities are brought forth by technical circumstances, it cannot be predicted on a technical basis which of them — if any — will become reality.

Given the lack of identifiable, direct mechanisms for the propagation of change across the different realms, utilities may operate under the implicit assumption that these realms are independent. As a result, effects across realms when they do occur could result in surprises. For example, utilities researching distribution automation today assume that they will continue to be the sole owners and operators of the T&D system in their territory. The T&D infrastructure is widely recognized as a natural monopoly and it is assumed that only the utility has both sufficient information and motivation to implement a distribution automation strategy that optimizes system performance.

However, these assumptions are based more on historical practice than assessments of present and future technical possibilities. It is technically feasible that generation facilities, sections of T&D systems and decentralized dispatching would be installed, operated and coordinated by independent actors rather than a single utility. A variety of business arrangements among such independent suppliers of "grid services" and their customers can be imagined. Thus, it is conceivable that utilities could develop distribution automation technologies for goals defined within the historical system structure only to discover that they have provided the technical basis for breaking their T&D monopoly.

Biases Against Supple Technologies

While technological change can have implications for system characteristics within the institutional realms, the reverse can also be true.

Specifically, an established technical system may exhibit "momentum,"[11] or resistance to the adoption of new technologies and operating practices that seem inconsistent with its current institutional structure or philosophy. In particular, the implementation of renewable energy technologies has been inhibited by institutional and cultural factors beyond simply cost and politics (such as subsidization of traditional fuels). These factors include accepted methods of cost and value analysis, the decision-making framework for additions or modifications to the physical system, and the prevalent perceptions about performance and technical compatibility of various technologies.

Classical evaluation methodologies for generation capacity additions were developed for large-scale, fossil-fuel based steam generation. Naturally, they do not lend themselves as tools for capturing and quantifying differences between options with diverse scales, siting potential, and output profiles. Specifically, traditional methods do not explicitly evaluate risk, project lead times, modularity, or environmental effects. They also fail to analyze the value of intermittent power output according to its coincidence with demand. In addition, the use of market discount rates penalizes capital-intensive technologies in a somewhat arbitrary manner. All of these factors tend to weigh against supple generation technologies.

The context in which options are evaluated is as important as the variables that are included. Traditionally, generation investments are considered separately from T&D investments and expenses and their value is assessed on a systemwide rather than a local basis. Both factors obscure the potential benefits of modular generation for the T&D system. Furthermore, traditional utility planning has been shaped by expectations of high demand growth and high reliability standards. Until recently, generation planning could be summarized as the problem of procuring enough supply to meet the ever-growing demand with a sufficient safety margin. This conceptual framework does not allow for fine-tuning supply and demand with smaller generation options and demand-side management.

Technology choices can also be affected by the working culture within organizations implementing them. For example, many power distribution engineers today prefer manually operated, mechanical switches to solid-state switches because "with a solid-state switch, you can't see whether it's open or closed."[12] For generation and T&D technologies alike, there may be a cultural gap between those designing and testing new gadgetry in research and development laboratories and those working "in the field," who have acquired experience with and thus a personal investment in existing technologies and ways of doing things. These differences will be reflected in their identification of weaknesses in the system and their assessment of corrective means, including new technologies.

One stumbling block for supple generation technologies has in fact been the lack of confidence on the part of utility engineers. Until PURPA mandated the admission of non-utility generators into the grid, it was widely believed that dispersed and intermittent generation would seriously threaten safety, reliability and power quality, particularly if it was not utility-owned. Technological improvements (as in the case of inverters), appropriate legal arrangements (e.g. to assure safe disconnection of generators from the grid when necessary), and several years of operating experience are contributing to a gradual process of building trust toward heterogeneous power generation within a grid. However, some deep suspicions still remain. For example, the operation of disconnected areas within a power system (or "islanding") is technically feasible and its use could help realize certain benefits of dispersed generation for service reliability. Nevertheless, it is prohibited in current utility practice.

Finally, technology choices are affected by cultural variables beyond the organizational setting. Many people still believe, for example, that renewable resources cannot make a significant energy contribution on a national or global scale because resources are insufficient, not accessible enough, or "solar energy just isn't strong enough for industrial applications."[13] If such beliefs are sufficiently pervasive in the cultural environment (though they might be unrealistic), they can surely stifle the enthusiasm of energy planners to consider and implement renewables.

All of these biases, embedded in the formal evaluation frameworks as well as the informal professional judgment of engineers and managers in the industry, have undoubtedly impeded the integration of supple technologies into utility systems so far. They constitute direct and indirect mechanisms by which an energy "path" defends itself against the challenges by alternative approaches. The question that remains is whether this cultural context will gradually yield to the pressure of technological reform from within, or whether the "softening" of energy systems would ultimately require a severe break from the existing culture.

Conclusion

Historically, dispersed power generation and electric utility grids have been considered mutually exclusive approaches to energy supply.

Lovins described the "deep structural and conceptual dichotomy" between the hard and soft path as follows:

> They are not *technically* incompatible: in principle, nuclear power stations and solar collectors could coexist. But they are *culturally* incompatible: each path entails a certain evolution of social values and perceptions that makes the other kind of world harder to imagine. The two paths are *institutionally* antagonistic: the policy actions, institutions, and political commitments required for each (especially the hard path) would seriously inhibit the other. And they are *logistically* competitive: every dollar, every bit of sweat and technical talent, every barrel of irreplaceable oil that we devote to the very demanding high technologies is a resource that we cannot use to pursue the elements of a soft path urgently enough to make them work together properly.[14]

Lovins thus saw the implementation of soft technologies as contingent upon a larger, explicit social commitment, while discounting the possibility of a limited but meaningful integration of soft technologies within the hard path.

A different view but one that also implicitly considers the hard and soft paths exclusive is widely held in the electric utility and engineering communities. Adherents to this view doubt the *technical* compatibility of soft technologies with a hard power system. They doubt this not necessarily in the sense that it could not be made to work at all but in the sense that it could not work well or efficiently enough to represent an economically viable option. To the extent that these beliefs are not justified by factual evidence, they too constitute a type of cultural incompatibility.

Some of Lovins' critics have contested the desirability of the soft path based on concerns about either the reliability of energy supply afforded by renewable resources or the anticipated social implications of such a conversion. Rossin, for example, argued that the conversion to the soft path would imply energy shortages and rationing while the existence of a reliable, traditional grid preserves options to install alternative generation equipment in niche applications.[15] Others have categorically denied the social and cultural dimensions of soft technologies, asserting that they would simply be implemented on the basis of rational decisions if and when their technical and economic characteristics were desirable.[16] None of the above interpretations is fully satisfactory in light of the current developments in the electric utility industry.

Practically all major industrial societies have been implicitly or explicitly committed to the path of ever-growing, large-scale, centralized power production for many decades. But as it turns out, this path has not evolved in quite the way Lovins foresaw. There are several reasons why the hard path has not been carried to its logical extreme in the U.S. and some Western European countries.

First, demand growth simply did not continue along the expected exponential trend and eroded the motivation for large-scale efforts to add generation. Second, the nuclear fission programs everywhere except in France have encountered enough public opposition to jeopardize investments, discourage utilities from relying on the nuclear resource, and prevent whole-hearted commitments to nuclear energy on the national level. Third, "hard path" construction and operating costs began to increase rather than decrease as they had for three quarters of a century. Fourth, transmission and distribution constraints became more significant. The weaknesses of the hard path have made those implementing it more receptive to soft or supple options — to the extent that it is arguable whether the power systems existing in the U.S. and Europe today can still be characterized as purely hard.

Lovins' soft path had a teleological character as in "deciding where we want to go and then taking steps toward that goal."[17] In his framework, soft or supple technologies would not be used *unless* society had made such a commitment. Today, however, it appears that supple technologies are being implemented without explicit social goals but as a result of internal economic motivation on the part of electric utilities. In the U.S., the PURPA legislation represented externally imposed pressure on the industry toward specific social goals and was obviously crucial for permitting this development in the first place. Since that initial impetus, however, the response of U.S. utilities to supple technologies is taking on an internal dynamic of its own. In this dynamic, supple options are recognized as marginally beneficial additions to or modifications of systems that are evolving qualitatively more than quantitatively as they seek to maximize efficiency and minimize costs. By means of this dynamic, a system which was once hard can incrementally take on soft characteristics.

The sum of incremental changes over time can be viewed as an adiabatic process in the thermodynamic sense of being reversible at any point along the way. Each incremental change is in principle reversible because the introduction of a new supple component does not require the system structure to change *a priori*. However, since the supple components *permit* structural change, it is possible that their sum over time results in an irreversible transformation across different realms of the system. If so, this result may be unanticipated even by those implementing and encouraging the supple modifications.

Even when the use of supple technologies is motivated internally by system optimization considerations, the cultural impediments to their implementation remain very real. How these conflicting pressures are resolved in specific instances will be an interesting subject for future study. In general, the notions of conflicting cultural and technical pressures and reversibility raise the question of how far the reconfiguration of power systems by incremental change can and will proceed or whether stable equilibria may be found in the spectrum between hard and soft.

The specific interactions between the various technical and institutional realms will be crucial to the question of system reconfiguration and require further study. The evidence suggests that while supple technologies open the door to a host of possibilities for reconfiguration and can thus be seen as necessary conditions for certain changes within the institutional realms, their existence and availability are not sufficient conditions for — or unambiguous predictors of — such a change.

The dynamic described in this chapter represents a mechanism for significant change, though it is not yet clear how this process will fit into the theory of the evolution of large technical systems. Since it represents a generative process that does not necessarily require quantitative growth or imply atrophy and elimination, the type of reconfiguration suggested here might be seen as a differentiation process analogous to the qualitative development observed in biological systems upon reaching maturity in terms of size or scale. It will be interesting to investigate whether evidence of similar phenomena can be observed in other technical systems.

Notes

1. Lovins, 1977, p. 38.
2. Lovins, 1977, pp. 38ff.
3. See Hirsh, 1989.
4. Wu, 1984.
5. Pacific Gas and Electric, for example, is installing a photovoltaic power plant on the customer side of a substation transformer that has been experiencing overloads on summer afternoons. The 500 kilowatt PV plant will enable PG&E to defer the upgrading of this transformer for an estimated five years. The economic value of this deferral is comparable in magnitude to the value of the plant's entire annual energy production (see Shugar et al., 1992).
6. Approximately five percent of the electricity generated in the U.S. is lost on transmission lines.
7. Nancy Rader, after Cogeneration/Small Power Production Quarterly Reports filed by California electric utilities (2nd Quarter 1991).

8. See Weyman-Jones, 1989; Roberts, Elliott & Houghton, 1991; and Ruff, 1992.

9. Hughes, for example, refers to "post-modern management styles" associated with less hierarchically structured technical systems (Thomas P. Hughes, "Modernism and Postmodernism," Presentation at the Energy and Resources Colloquium, April 1992, University of California, Berkeley).

10. James Eyer, PG&E, personal communication.

11. Hughes, 1983, 1987.

12. Jimmie Yee, PG&E, personal communication.

13. As expressed by an innocent geophysicist.

14. Lovins, 1982, p. 552.

15. Rossin, 1982.

16. See, for example, Blackman and others in Nash, 1979.

17. Lovins, 1977, p. 49.

References

Daneke, Gregory A. 1979. "Toward an Alternative Energy Future." In *New Dimensions to Energy Policy*, ed. Robert Lawrence. Lexington, Mass.: D.C. Heath and Company.

Gönen, Turan. 1986. *Electric Power Distribution System Engineering*. New York: Mc-Graw-Hill.

Hirsh, Richard F. 1989. *Technology and Transformation in the Electric Utility Industry*. Cambridge: Cambridge University Press.

Houston, Douglas A. 1992. "User Ownership of Electric Transmission Grids." *Regulation* (Winter 1992): 48-57.

Hughes, Thomas P. 1983. *Networks of Power*. Baltimore: Johns Hopkins University Press.

_____. 1987. "The Evolution of Large Technological Systems." In *The Social Construction of Technological Systems*, ed. Wiebe E. Bijker, Thomas P. Hughes, and Trevor Pinch. Cambridge: MIT Press.

Kahn, Edward. 1988. *Structural Evolution of the Electric Utility Industry*. Lawrence Berkeley Laboratory Report LBL-26165.

Lovins, Amory B. 1976. "Energy Strategy: The Road Not Taken?" *Foreign Affairs* (October 1976): 65-96.

_____. 1977. *Soft Energy Paths*. San Francisco: Friends of the Earth.

_____. 1982. "Energy Strategy: The Industry, Small Business, and Public Stakes." In *Perspectives on Energy*, ed. Lon C. Ruedisili and Morris W. Firebaugh. New York: Oxford University Press.

Nash, Hugh, ed. 1979. *The Energy Controversy. Soft Path Questions and Answers*. San Francisco: Friends of the Earth.

Natoli, Donald R. and Wu, Felix F. 1983. *A Distribution Automation Approach to Controlling Dispersed Storage and Generation Systems*. Universitywide Energy Research Group California Energy Studies Report UER-116. University of California.

Roberts, Jane, David Elliott and Trevor Houghton. 1991. *Privatising Electricity. The Politics of Power.* London: Belhaven Press.

Rossin, A. David. 1982. "The Soft Energy Path: Where Does It Really Lead?" In *Perspectives on Energy,* ed. Lon C. Ruedisili and Morris W. Firebaugh. New York: Oxford University Press.

Ruff, Larry E.. 1992. "The U.K. Electricity Market: The Experiment and its Lessons." Presented at Program On Workable Energy Regulation Conference, Sacramento, April 1992.

Russell, M.C. and Kern, E.C. Jr. 1991. *Experiences and Lessons Learned With Residential Photovoltaic Systems.* Palo Alto: Electric Power Research Institute, GS-7227 Research Project 1607-15.

Shugar, Daniel S., Mohamed El-Gasseir, Allen Jones, Ren Orans, and Alexandra Suchard (von Meier). 1992. *Benefits of Distributed Generation in PG&E's T&D System: A Case Study of Photovoltaics Serving Kerman Substation.* PG&E Research and Development Report. San Francisco: Pacific Gas and Electric Co.

Summerton, Jane and Bradshaw, Ted K. 1991. "Toward a Dispersed Electrical System: Challenges to the Grid." *Energy Policy* (Jan/Feb): 24-34.

Weyman-Jones, Thomas G. 1989. *Electricity Privatisation.* AMERG economic studies 2. Brookfield, VT: Avebury/Gower Publishing Company.

Wu, Felix F. 1984. "Stability, Security, and Reliability of Interconnected Power Systems." *Large Scale Systems 7.*

11

Broken Plowshare: System Failure and the Nuclear Power Industry

Gene I. Rochlin

My name is Ozymandias, king of kings:
Look on my works, ye Mighty, and despair!"
Nothing beside remains. Round the decay
Of that colossal wreck, boundless and bare
The lone and level sands stretch far away.

(Percy Bysshe Shelley, *"Ozymandias"*)

Part One:
Introduction

Nuclear power represents a new class of large-scale systems based on modern technologies that do not fit the historical model of adaptive and conformal growth to scale. Comprehensive system design for delivery of services *at full scale* was not only part of the original conception of nuclear power but in fact the dominant consideration, the result of a desire to capture social and political gains by exploiting a new scientific capability. Design and growth were not driven by technological development but for the purpose of putting the system into place.

From the outset, commercial nuclear power was intended to be a large-scale technical system with varied and closely-linked components. The designers never envisioned individual, isolated, unique reactors embedded in socio-political controversy but instead saw an extended *system* of nuclear fuel production, nuclear waste management, plutonium recycle and technical evolution in which current reactors were to be only a first step towards an integrated nuclear power system.

Commercial nuclear power is therefore more than just a series of nuclear plants. As designed, it is a complex techno-social-industrial system configured to prior expectations about structure and linkages. Radical reconfiguration has occurred if the nuclear power *system as deployed* now differs greatly in its parameters and system-environment interactions from what was originally envisioned. Commercial nuclear power is perhaps the most interesting example because there is more than a single case. If success is defined in terms of deployment of the complete cycle as planned, there are two cases of clear success (in France and possibly Japan), several cases of artifactual success combined with systemic failure (in the United States, Germany and possibly Sweden), and some cases where system failure occurred early enough to prevent the further deployment of artifact and instrumentality (in Switzerland and perhaps Sweden).

Taking the case of Japan aside for lack of comparable data, the outstanding case is France. If the failure of nuclear power represents the evolution of modern Euro-American societies, the questions to be addressed are: why is France different and why was nuclear power so readily deployed there?

France is the only democratic European country in which the envisioned large-scale system was successfully deployed. In contrast, the nuclear industries of the U.S. and Germany are cases of radical reconfiguration brought about largely by a change in the exogenous system environment. Whatever the future of nuclear power in these countries, present configurations of nuclear power plants do not constitute elements of the envisioned systems. Indeed, they can hardly be described as systems at all; they might better be viewed as the techno-industrial residue of the rapid collapse of a postulated and once-growing large-scale technical system.

I argue that successful deployment of nuclear power in France was a direct consequence of political centralization, technocracy, hierarchy and social isolation. This led to an unusual degree of acceptance by governmental authority, mediated by scientific and technical elites. It was supported by coordination with the deployment of a nationalized electrical utility and a centralized nationwide grid that was designed to integrate the nuclear plants as they deployed.

Part Two:
A Brief History of Commercial Nuclear Power

"Power too Cheap to Meter."

(Then AEC chairman Lewis Strauss at a function in New York, September 16, 1954)

During World War II, the U.S. nuclear weapons program drew heavily on the expertise of its allies, particularly Britain, Canada and France. After the war the U.S. locked up its technology, classifying the Manhattan Project materials so secret that even foreign scientists who had contributed the most to the war effort could not look at their own notes and experimental results. The Atomic Energy Act of 1946, which also created the U.S. Atomic Energy Commission (AEC), solidified as national policy the United States' intent to focus its atomic research and resources on weapons and naval propulsion but not on commercial power reactors.[1] The U.S. held its atomic secrets tightly for eight years, in effect shutting out its own commercial firms from new scientific and engineering developments.

In Britain, Canada, France and Sweden, the histories were quite different. Canada eventually opted to forego nuclear weapons and focused on a reactor design independent of the now-blocked war materials.[2] Britain and France, intent on pursuing weapons as well as power, designed dual purpose systems based on the natural-uranium, gas-graphite reactor that produced both plutonium and electricity.

In 1950, with its allies beginning to explore the commercial use of power reactors, the U.S. grew concerned that it would lose its leadership on the peaceful side of atomic energy to its former allies or the Russians. Despite protests from Congress and AEC scientists, the AEC in 1951-1952 moved towards allowing industrial participation in its own program, issuing some security clearances to private vendors and creating a study group on dual-purpose reactors. A few utilities indicated that they might be willing to commit a few tens of millions of dollars for a commercial venture — less than a tenth of the estimated costs.[3]

In 1953 came Eisenhower's famous "Atoms for Peace" proposal, followed by the 1954 revision of the U.S. Atomic Energy Act and the 1955 Geneva Conference. The U.S. sharply reversed its course. Now the commercial uses of atomic power were to be promoted, at home and abroad. Moreover, the U.S. AEC was to take an active role not only in disseminating the technology, but in nurturing the infant industry until it acquired the skill and experience to compete on its own. Stimulated by the U.S. challenge, Britain, France and Canada were then to move even faster. Sweden had just started up its first reactor and had formed its first long-term commercial nuclear power strategy.

In the U.S. with its tradition of private enterprise, anti-socialism and anti-corporatism, the Atomic Energy Act of 1954 not only encouraged but mandated a separation between civil and military uses and private and government projects. On the other hand, in Britain, France and Sweden such a separation was seen as not only unwarranted but inefficient. Thus began the first of many divergent nuclear paths.

Power Plants and the Nuclear Fuel Cycle

Whereas France and the U.K. deployed dual-purpose power and plutonium reactors, the United States began a series of government programs designed to promote fully commercial reactor designs independent of its weapons program.[4] In 1955 and 1957, the U.S. announced new rounds of a power reactor demonstration program to encourage private industry to draw upon AEC expertise to develop new ideas and designs for commercial power reactors. Meanwhile, Britain, France and Canada were deploying their first commercial designs.

Three major technical factors also clearly separated the U.S. from the rest. First, the U.S. had a surplus of capacity for the difficult and secret gaseous diffusion technology for enriching uranium. Second, the nuclear submarine program had already produced a light-water cooled and moderated reactor (a Westinghouse design) that was easily adaptable for commercial uses. These two factors combined to short-circuit the AEC's announced program for competition among multiple designs, so that the light-water reactor (LWR) quickly emerged as the dominant technology in the American system.[5]

The third factor, namely closure of the nuclear fuel cycle, had the most significant impact on the future of commercial nuclear power as a large-scale system. A nuclear power reactor is one element of a network of facilities and activities that make up the complete nuclear fuel cycle. A LWR fuel cycle requires facilities for mining the ore, extracting the uranium, changing it into gaseous form for enrichment, enriching it, fabricating fuel elements, managing the spent fuel discharged from the reactor and disposing of the radioactive wastes.[6] Because of the proliferation risk and radiation dangers, the complete cycle also includes numerous socio-technical activities for safeguarding, protecting and accounting for various materials every time they are shipped or handled.

A central part of the cycle is devoted to separating and recovering plutonium that is generated in natural or low-enrichment uranium fuel in the reactor. The crucial step is chemical reprocessing of the spent fuel to separate the plutonium. From the perspective of the nuclear industry, spent fuel is not a waste but the repository of an unrecovered nuclear resource. Only after the plutonium is extracted are the residual radioactive actinides and fission products in the spent fuel considered to be wastes — usually referred to as high-level wastes (HLW) because of their intense radioactivity.

The difference between spent fuel and HLW has continued to confound the commercial industry for decades. In military fuel cycles, it is the plutonium — not the energy — that is sought. HLW is the residue of plutonium separation. In the original commercial fuel cycles used by the French, British and Soviets, the plutonium for military programs was extracted as a matter of course, so that the nuclear fuel cycle was complete. The government and not private industry was to be responsible for the extraction, security and protection of weapons-usable plutonium and the management and disposal of high-level nuclear wastes.

The American nuclear industry and AEC shared the same original vision, except that U.S. law prohibited diversion of commercial reactor plutonium for military uses. In the U.S., private industry was to be responsible for fuel reprocessing, plutonium extraction and recycling (for purely *commercial* purposes) and HLW management and disposal. The U.S. industry was hard pressed in its early years to find adequate uses and designs even for uranium-fueled reactors. No commercial reprocessing facilities were built and no plans for plutonium recycle existed.

Although plutonium use was not actually *necessary*, proponents of nuclear power had already suggested closing the fuel cycle through plutonium recycle — not only for technical efficiency but also for ideological reasons. Achieving a closed plutonium cycle is viewed by some as the single most important indicator of success in a nuclear industry. But since the industry and the AEC viewed the nuclear fuel cycle as an entirely closed cycle, they could never allow planning for the disposal of spent fuel as a waste. This left the cycle's back end in limbo.[7]

The international community of nuclear scientists and engineers is among the most internationally socialized in the world. It is, therefore, not surprising that as the commercial nuclear industries moved from the drawing boards to reality during the 1960s, they shared a common view of what constituted a commercial nuclear power system. It was never intended to be a collection of individual plants operating in relative isolation — let alone the odd assortment of hand-crafted units that characterizes the U.S. and to some extent Germany. It was to be a large-scale, integrated system with elaborate interconnections and material flows.

Since uranium resources were perceived to be finite, the reactor designs of that period were viewed as interim. In the long-run, the nuclear fuel cycle was to become essentially self-sustaining through the fast-breeder reactor (FBR), which produced more plutonium than it consumed and expanded the energy value of raw uranium fuel some fiftyfold. Thus, fuel-cycle closure was a necessary — not a contingent — goal.

The system's major socio-political flaw was that it had been constructed by a group of engineers and scientists who were relatively unfamiliar with and insensitive to the social and organizational demands of large-scale systems. They failed to consider the reluctance of electrical utilities to move into the new technology. After decades of regulation, utilities were conservative about ideas and investments in an era of cheap and relatively abundant energy supplies. Electrical rates were falling continuously and were expected to continue to drop. In such a climate, nuclear power looked like a risk with little benefit.

Part Three:
Commercialization

"Tout électrique! Tout nucléaire!"

("All electric! All nuclear!" — EdF's slogan in 1971 during its 25th anniversary year.)

The United States

In the early 1960s, the U.S. was still trying to get a design commercialized. Westinghouse had a promising redesign of its submarine reactor, the pressurized-water reactor (PWR); General Electric had a design that was similar conceptually but different enough in detail to escape Westinghouse patents — the boiling-water reactor (BWR). In 1962, the AEC abandoned the multiple technology approach to push harder for commercialization of existing designs. This led directly to the "great bandwagon market" of 1963 after GE sold its first complete turnkey plant, Oyster Creek. The plant was sold at a loss to stimulate the business, although that was not apparent then. As other turnkey plants were sold at loss-leader prices, energy projections soared, feeding back into the nuclear optimism.

Undaunted by constraints of social or political reality, the AEC predicted that as many as 1,000 plants would be operating by the year 2000. But the bandwagon had already crashed and the future of commercial nuclear energy in the U.S. was in serious doubt. There were many reasons, some completely technical, some socio-technical and some having to do with public reaction and resistance to the system's operation at the anticipated large scale.

On the technical side, the first commercial plutonium-breeder reactor (the Enrico Fermi plant outside of Detroit) suffered a near-meltdown in 1957, bringing technical experimentation to a near halt. Without a breeder reactor, the civil market for plutonium would be quite limited. Furthermore, the AEC's first attempts at siting a repository for high-level nuclear wastes turned into a dismal failure, dealing the program the first of many setbacks from which it has not yet recovered. Thus, two key elements of system closure were put on hold.

The industry also had serious internal problems. The first LWR were oversold and under-priced. Firms engaged in a reckless round of scaling up reactor size and power without first gaining operating experience. Remarkably, this technically optimistic approach that characterized the new breed of utility owners, operators and their suppliers was mirrored in turbine size and systems integration, with negative effects for those areas as well.[8]

Meanwhile, two anti-nuclear currents were growing. The environmental movement in the 1960s brought the first organized opposition derived from linkage to the nuclear weapons program and nuclear testing, increased fears of radiation as a carcinogen, and the AEC and industry's public failures (mainly the waste repository fiasco and Fermi accident). The second anti-nuclear current arose from the perceived negative externalities of operating the system as a closed system as originally intended. The social aspects of radioactive waste management — safety, repository management and the long life of wastes — were beginning to be better understood. More important, however, were the consequences of separating and recycling plutonium in the fuel. The Nuclear Non-Proliferation Treaty of 1968 had brought to the general public a better understanding of why the theft of plutonium fuels or the production of plutonium in reactors greatly increases proliferation risks compared to a low-enrichment open cycle. The means and methods required to adequately guard and secure the plutonium during shipment and use would also create serious intrusions on civil liberties and rights.

By the late 1960s and early 1970s, as a result of controversies and delays, the U.S. had thus not moved towards the closed nuclear fuel cycle. The environmental movement was in full swing and opposition to nuclear power was growing rapidly. The nuclear industry might have negotiated the narrow channel successfully, but a series of dramatic program failures by the AEC in the 1970s effectively closed off the possibility of a non-controversial outcome. The AEC and Congress accepted private industry's argument that the government should be responsible for the disposal of high-level nuclear wastes. But the chosen site (a Kansas salt bed) proved unsuitable, and the resulting wave of bad publicity, poorly handled by the AEC, resulted in a loss of credibility for the U.S. waste management program.

In 1974, the old AEC was disbanded and its parts redistributed. The realignment was also used to separate its regulatory arm into the independent Nuclear Regulatory Commission (NRC) on the grounds that a single agency should not promote and regulate nuclear power. No longer could a member of Congress base a career on mastering the intricacies of the nuclear fuel cycle. Public resistance to nuclear facilities was now compounded by increasing difficulty in obtaining favorable legislation.

By the mid-1970s, industry thought the situation in the U.S. was unsatisfactory. There was no repository or civil facility for nuclear fuel reprocessing and plutonium's extraction and recycling. To this end, the AEC was able to induce one private vendor to build a commercial reprocessing plant. Under the United States' new environmental mandate, the NRC prepared a Generic Environmental Impact Statement (GESMO)[9] that immediately became embroiled in controversy.

For the first time, there was serious public discussion of the negative impacts of a complete nuclear fuel cycle, *even if there were no accidents or other unseemly events*. The issues included the security of nuclear waste shipments, the issue of plutonium separation and storage facilities, the civil liberties' implications of safeguarding a large-scale plutonium fuel cycle, and nuclear proliferation and terrorism.

These issues were never resolved. President Carter's declaration that the U.S. would not pursue a plutonium-based fuel cycle resulted in a withdrawal of the GESMO and a sudden termination of the public hearings. The U.S. was then in a position in which the key element of the nuclear power system was declared to be against government policy. The waste program was once again thrown into disarray over the definition of terms — a disarray that was to turn to complete disorder when newly-elected President Reagan threw out almost all of the planning for a civil waste management program that the Carter Administration had developed.

The U.S. experienced two high points in nuclear power plant orders: the great bandwagon market lasted until 1967-1968, followed by consolidation and another optimistic wave in 1971-1974. By that time, the growth of the environmental movement had resulted in sufficient public controversy to essentially cancel the pro-nuclear effects of the first oil crisis in 1973-1974. The controversy over nuclear power did not take place until 1976-1980, reaching its peak in 1979 with the accident at Three Mile Island (TMI) adding to the ongoing debates about waste management, the use of plutonium and the problem of plant siting.

Over the years, the public focus on the safety of nuclear power plants has drawn attention away from the resulting systems failure. It is clear that the systems-level failure was starting even before the controversy reached the heights of polarization. By the late 1970s when the public debate peaked, the systems failure was already almost complete; the accidents at TMI and Chernobyl only drove additional nails into the coffin.

Little has changed since the 1970s in terms of power plant orders, plutonium facilities, waste management, financing, social and industrial structure, or governmental intervention and support. What the U.S. now has may seem large scale in absolute terms. It has the most nuclear power capacity in the world. One hundred and ten reactors produce about twenty percent of the U.S. electricity demand. But the reactors are a hodge-podge of independent designs of different ages and capacities, operated as individual plants or in small groups without any systemic integration or standardization. There is no viable plutonium-breeder reactor program or fuel reprocessing. Nor are there facilities for plutonium storage, fuel fabrication or shipment. There is no large scale central facility for storing spent nuclear fuel; there is still no high-level nuclear waste repository nor one on the horizon.

Nuclear advocates point out the number of plants and their overall operating records as positive signs and criticize public and governmental bodies alike for an anti-nuclear bias. They react strongly against words like failure and in some sense they are right. The system that was supposed to have been put in place has not failed. It has simply disintegrated, leaving the existing plants as isolated islands of technical prowess deprived of their original systemic context.

France

Unlike the British, for whom the gas-graphite reactor was part of a deliberate strategy to co-develop commercial technology and produce weapons material, the French program was ostensibly begun for the primary purpose of developing power reactors. Left in a state of devastation by the German occupation and the fighting of 1944-1945, France embarked on a centralized policy of modernization and economic development, putting its energy industries in the exclusive hands of the state and empowering the engineers of the *grands corps de l'état* to lead them. To foster this development, private electrical utility companies were nationalized in 1946 and placed into a single, state-run utility, Electricité de France (EdF).

The French equivalent to the U.S. AEC, the Commissariat à l'Energie Atomique (CEA), created in 1945, was granted autonomy from political control and explicit freedom of action. Although the CEA mandate included national defense, the question of building an atomic bomb did not arise at first. Both technical and industrial resources were still scarce and there was neither the ample wartime experience nor the strong political motivation that drove Britain. The turning point came in 1951 when Gaullist deputy and newly appointed Secretary of State Félix Gaillard became a member of the CEA steering committee. Gaillard advocated short-term benefits rather than long-term research, pushing the CEA to opt for a full-scale demonstration reactor.

Lacking uranium enrichment technology, the CEA opted for a natural uranium design, settling on gas-graphite on political, technical and economic grounds; the production of usable weapons-grade plutonium was seen as a side benefit. Even though the CEA did not intend to develop a weapons program, it wanted to keep the option. It also specified the construction of facilities for extracting and storing plutonium. Since France also had its own reserves of uranium, the first design was for a complete and entirely national fuel cycle, lacking only a specific method and location for disposal of the wastes. Five years later both the demonstration gas-graphite reactor (G1) and the plutonium reprocessing plant were fully operational at the CEA site at Marcoule.

As in the U.S., the original fuel cycle was an interim design with the goal of producing a closed breeder cycle running on plutonium. The motivation for plutonium recycling was to train technicians, acquire technical experience, and create the stockpile of plutonium necessary for experimentation and design of a breeder reactor.[10] The ambiguity of Marcoule's purpose was also useful since it could be represented as being civil, military or somewhere in between according to the political climate.

As the CEA began to focus increasingly on the new weapons program, it fell to EdF to develop the commercial side of the French nuclear industry. Because EdF had not been involved in the original decision, there was at first no plan for electricity production from the demonstration gas-graphite reactor. To resolve the tension between EdF and CEA as to who would supply nuclear-generated electricity to France, EdF was invited to add a small generator to its gas-graphite reactor and was involved in the design of the second one. However, the CEA selected and directed the industrial firms involved.

In 1955, EdF and CEA began a commercial gas-graphite reactor program designed to generate 800 megawatts (MW) of electricity by 1965. Oversight was provided by a new and separate advisory council of industrial and governmental representatives (including CEA and EdF), and the Commission Consultative pour la Production d'Electricité d'Origine Nucleaire (PEON). At first, CEA was to provide the nuclear portions and EdF the steam generating ones. This division of labor was a source of inter-agency tensions that EdF resolved through the process of design and specification. The 60 MW EdF 1 plant at Chinon (1956) was optimized for efficiency rather than plutonium production and was less independent of non-French technologies and innovations.

With EdF's success at Chinon, commercial nuclear power expanded in France. Because both EdF and CEA viewed nuclear power as part of a comprehensive system that included plutonium extraction, recycling and breeders at maturity, EdF reactors were embedded in a growing large-scale system from the beginning. Many at EdF regarded the choice of gas-graphite as an unfortunate consequence of CEA's continuing search for a parallel weapons program and they continued to search for an alternative interim design to carry the industry through the breeder reactor's developmental years.

In 1969 and 1973, PEON recommended that France use LWRs and license U.S. technology. Because CEA had developed enrichment capabilities for military purposes, the fuel cycle could now be closed entirely under French control. With military reactors producing adequate plutonium, the gas-graphite program was then terminated. The new LWR program began slowly with two 900 MW PWRs at Fessenheim and four BWRs at Bugey.

After the oil shock of 1973-1974, France opted to expand its nuclear program, choosing Westinghouse's PWR design. The variance from the U.S. case is remarkable. The CEA/EdF alliance retained central control. The 900 MW design was standardized; Framatome was assigned to become the single reactor producer with Alsthom-Atlantique building the plant's non-nuclear parts. Nuclear power became a centerpiece of planning. Furthermore, the five-year plans were augmented with multi-year contracts which were legally binding, making the nuclear system one of the most *dirigisite* (or, very roughly translated, inflexible). Other entities created were Cogema (Compagnie Générale des Matières Nucléaires) formed in 1976 as a private subsidiary of the CEA for fuel reprocessing and plutonium handling, Novatome to build and operate the fast-breeders, and other companies for mining, milling, waste disposal and related activities.

The LWR was implemented in France as an element of a complete fuel cycle. As in the U.S., it was seen as transitional in the long term.

Unlike in the United States, the elements for its successor were designed and implemented from the outset. The French fuel cycle looks very much like the original U.S. intention. Yet France was able to implement and deploy its system at full scale — 75 percent of French electrical power at this point. During the growth period, the French altered the Westinghouse design to suit their needs and sent Framatome to compete worldwide against American and German reactor and nuclear fuel cycle vendors. Cogema also went to scale with many overseas contracts and the French began their own plutonium recycling and pushed their fast-breeder program.

Prior to 1974, there was hardly a nuclear debate in France at all. But in a partial spill-over from the German experience, large protests took place in 1976-1977, culminating in a riot at the Super-Phenix breeder site at Creys-Malville. After election losses by anti-nuclear groups in 1978, protest activity began to wane and the French government made no move to co-opt the protestors or bring their voice into decision-making. By the end of 1978, France had the only fully integrated nuclear corporation in the world. Eighteen 900 MW plants were to be built between 1974 and 1984 when Framatome was to move to an even larger standardized 1,300 MW plant.

By the late 1980s, matters began to change. Energy conservation and economic recession combined to limit the growth of the electrical system, putting a crimp into French-planned EdF expansion — a crimp tightened by the public response to the Chernobyl accident in 1986 and the shut-down of Super-Phenix for repairs in 1987. The Mitterand program of decentralization, started in the early 1980s, began to affect even the technocrats of the Institute Polytechnique, with nuclear planning and regulation becoming a more open process. Nevertheless, France had succeeded in bringing its nuclear power system irreversibly to scale. The triple dynamics of joint CEA/EdF cooperation, central planning and long-term contracts, and the role of EdF as the single generator, distributor and transmitter of electrical power were important to the process.

Unlike the U.S. with its patchwork of utilities, EdF was a national utility whose scale and overall capabilities were appropriate to the planning and technical demands of nuclear technology. Moreover, as a state within a state it was able to develop, plan, expand, deploy and control the power grid to accommodate the construction and phase-in of the nuclear plants. It was aided by its ability to contract for sales of equipment, technology, and much exported electrical power.

The troika that pulled France's nuclear sled was Gaullism, *dirigisme* (i.e. state intervention based on positivistic planning) and centralism. Gaullist attitudes remain crucial because they foster social isolationism and the notion of an independent French path, sheltering the French industry from the growing international cycle of anti-nuclear criticism. *Dirigisme* encouraged the nuclear planners to look ahead five or ten years, enabling them to make decisions that looked socially, politically and financially risky in other countries. Centralism promoted standardization of reactors and technologies, controlling costs and eliminating the technical competitiveness that led the U.S. industry to expand reactor power and size without accumulated experience. Together, these three factors represent precisely the modalities of central, technocratic control which were necessary for the technologically rigid nuclear system to flourish without dislocation.

Germany

The origins of the program in the Federal Republic of Germany (formerly West Germany) differ greatly from those of the U.S. and France in that they were derived solely from commercial concerns. Lacking a military program, Germany never had a powerful, centralized agency such as the AEC or the CEA. When the country did form an Atomic Energy Commission in 1956, it was an open ministry with little concern for secrecy or military missions. Indeed, some of the technical leadership was deliberately diverted to research facilities at Jülich and Karlsruhe. Thus, the German program never followed the sequential logic of the U.S. or France and its ministries never achieved the power of the AEC or the CEA.

In most cases, analyzing the political economy of industrial policy begins with distinguishing between public/private and centralized/decentralized planning and control. Germany does not fit neatly into these descriptive boxes. In the polycentric country, grid and power plants are owned by the Länder (state governments) and private enterprise, with the further complication that political units may also own shares in the private companies. Power is likewise shared between Bonn, the Länder and private enterprise in a modified corporatist model of great complexity.

Germany had a deliberately weak central government, with a great deal of power granted to the Länder, and was considerably concerned about overly close connections between the central government and large technical industries. Although planning is done at the national level, the Länder control the resources as well as regulatory and licensing processes. It is difficult for the federal government to challenge — let alone over-ride — the Länder's policies and decisions. The structure is deliberately non-hierarchical. Although the federal government can act with authority, it cannot exercise centralized, authoritarian control in many areas relevant to commercial nuclear power. Moreover, nuclear responsibilities are not only fragmented among agencies but partitioned out between the government and contracted private firms.

As in the U.S., in its early phases the German plan encouraged many reactor designs. As in France, the rhetoric of the German program was a technocratic vision of progress through industrial technology. But institutional differences from its allies were great. The German Atomic Commission, formed in 1956, had representatives of the trade unions and political parties as well as government and industry. Despite demands for nationalization, the Atomic Energy Act of 1959 called for a corporate structure of private industries with heavy state subsidies.

The first commercial reactor order (the BWR at Gunderemmingen) was placed in 1962, at roughly the same time as Oyster Creek and with somewhat the same effect. The major competition during the exploratory period (1962-1970) was between AEG BWRs (GE license) and Siemens PWRs (Westinghouse license). Nevertheless, research continued not only on the fast-breeder reactor and plutonium separation facilities but also on other reactor designs. Germany had previously set a deliberate course of treating the nuclear sector as any other market activity, avoiding central planning as well as central control. By 1970, it had consolidated. AEG's problems led to the merger with the Siemens reactor-turbine divisions to form a single manufacturer, Kraftwerk Union (KWU). Since then, the Siemens-KWU team has dominated the German commercial reactor industry.

As in France, the steady growth of the nuclear industry in 1970-1975 changed to a program of intended rapid expansion after 1975. Indeed, the French program was sometimes cited as the role model. The maturing German industry had a standardized reactor design in the KWU PWR, a potentially growing export market, and an ambitious plutonium and breeder program. The superficial technical similarity to the situation in France masked an enormous difference in institutional and social structure.

Plans for a complete nuclear fuel cycle were formulated early, with the first demonstration plant at Karlsruhe ordered in 1966 in preparation for development of a much larger plant later, including some possibilities of cooperation with France and/or Great Britain. In 1972, an order was placed for a much larger unit at Kalkar. Plans for a plutonium reprocessing and fuel fabrication facility at Gorleben in Lower Saxony were announced in 1976 as part of a German effort to have a nationally closed fuel cycle.

The year 1976 also marked the beginning of the years of the most violent opposition to nuclear power in Germany, climaxing with a violent clash at Grohnde. The violence was to quiet down, but the controversy was not suppressed (as in France). Instead it gained political power within the Länder, the press and academic circles. As in the U.S., this resulted in a great slowdown in licensing procedures and new plant orders. It also brought the cancellation of the planned reprocessing plant at Gorleben and the reluctance of existing plants to contract for plutonium fuels abroad.

None of these factors, however, was as important as the strong social movement against the socio-political consequences of continued expansion. Once the ideological anti-nuclear left allied with those resisting the spread of techno-industrial complexes (such as those on the Ruhr), nuclear power became embroiled in the broader controversy over the social as well as industrial future of Germany. Ultimately, even the fast-breeder program at Kalkar was cancelled, leaving the fuel cycle open and in disarray. Although the German government has never explicitly eschewed plutonium use as the U.S. did, the situation in Germany is now much like that in the U.S.

Without a strong central plan or central leadership, a coherent national program, an integrated grid and an operating fuel cycle support system, the present German system is perhaps even more reconfigured than the American one. Moreover, regional and local variations in attitudes and capabilities differ more drastically and widely than in the U.S., and the distribution of plants is very uneven. Most of the plants built operate very well. But at present they are no more than that — individual nuclear plants for making electricity and not, as had been planned, the basic elements of a nuclear-based electrical system.

Sweden

While West Germany was institutionally similar to the U.S., Sweden was at first closer to France. Most of Sweden's nuclear scientists were closely tied and from the same school (Chalmers). As a result, Sweden had a miniature version of the French nuclear technocracy, although acceptance of expert judgment was more social than cultural. Policy was centralized, as was the industry. But unlike in France, industrial and policy processes were relatively open. In this regard, Sweden was a halfway case between the U.S. and France.

The Swedish Royal Commission on Nuclear Energy was formed in 1945, evolving to a permanent advisory atom committee with six scientists, four industrialists and two government (defense) representatives. Military aspects were at first deferred, but nuclear weapons were intensely studied from 1958 to 1968, when the government officially dropped military research and signed the Non-Proliferation Treaty.

The institutional structure was a mix of national and corporate. AB Atomenergi, formed in 1948, was responsible for fuel-cycle development and Sweden's first experimental reactor. In the 1960s, it set up a committee with the major electrical equipment manufacturer ASEA and the state power board, *Vattenfall*, for continuing exploration of commercial possibilities. It was decided that self-sufficiency in fuel was not to be a high priority. In the early 1960s, LWR as well as indigenous heavy water designs were explored.

By the 1960s, nuclear euphoria had diminished. Although Atomenergi worked on uranium enrichment and even reprocessing, these activities were not efficient for a small country that lacked a military program. The state power board's support for the heavy-water project had also declined and ASEA developed its own BWR which it sold to a consortium of private utilities. In 1966, a thirty-year agreement with the U.S. for the supply of enriched uranium fuel guaranteed both the future of the LWR in Sweden and the externalization of the supporting fuel cycle. The national government intervened to foster LWR development, taking over AB Atomenergi as a research agency and merging Atomenergi and ASEA to form ASEA-Atom (half private and half government-owned) to build and sell reactors. The "Swedish Path" that had led to exploration of a complete fuel cycle was essentially abandoned.[11] By the 1970s, Sweden was a marketer and consumer of reactors but not fuel cycle equipment or services.

Sweden is a mixed case. Unlike most industrialized countries, production and distribution of electricity are largely separated. Yet there is extensive public-private collaboration. Half of the utility capacity is nationalized, half is private. The major reactor vendor and builder was a joint government-industry consortium, and although ASEA-Atom designs and builds its own reactors, some are indigenous designs and some are licensed U.S. designs.

During the early 1970s, Sweden seemed destined for a nearly all-nuclear future. With twelve reactors on-line or ordered, it had the largest per-capita nuclear energy program in the world. But it was already embroiled in political conflict. The growing anti-nuclear movement in America had spread to Sweden, aided by close links between many dissident Swedish scientists and politicians and their American colleagues. When Prime Minister Fälldin was elected on an anti-nuclear platform in 1976, he was unable to overcome the techno-bureaucracy and found little general support from a public still strongly supportive of centralized decision-making. By 1979, however, with a Liberal government in office, the delicate balance of political forces was tipped by the Three Mile Island accident, leading to a 1980 national referendum calling for the phasing out of nuclear power by 2010. This situation remained stable until quite recently when increasing concern about environmental impacts and the prospect of greater power exports put the matter of phasing out up for discussion once again.

The 1970s debate permanently changed the Swedish situation by moving the nuclear issue into the center of the struggle for political power. Even in highly technocratic and non-oppositional Sweden, the matter of nuclear decision-making has been almost completely removed from the hands of the technocratic elite and into the public arena. But the outcome was foreshadowed by the earlier decision to abandon nuclear power as a system and therefore neglect the back end of the fuel cycle. Isolated in context and deprived of systemic relationships, the reactors were as vulnerable to criticism as they were in the U.S. and Germany.

Part Four:
Public Opinion and Public Response

The 1986 accident at Chernobyl was a political watershed for nuclear power everywhere and most notably in those European countries that suffered real or imagined harm from the fallout. But the success or failure of nuclear power *as a large-scale technical system* had already been determined by the mid-1980s. It is, therefore, the earlier period that must be analyzed to determine whether the growth of the environmental movement and the strident anti-nuclear public debate that followed was indeed a primary causal factor.

There was little *general* public opposition to nuclear power until around 1970. Many analysts attribute the subsequent growth in opposition to factors arising from the radical upheavals of the late 1960s: student uprisings, the Vietnam war, the growth of the environmental movement and the legacy of the counter-culture. These analyses only show, however, that such factors are covariant; they do not address the root causes for the polarization of the nuclear debate or the ideological depth of the anti-nuclear commitment.[12]

The oppositional movement to nuclear power was the manifestation — and not the cause — of the reconfiguration of the systems in the U.S. and Germany, and probably in Sweden as well. There was a growing disjuncture between the rigid system requirements and the direction and degree of the social, cultural, and political milieus in which they were to be deployed. However, there is supporting evidence that the major evolutions in public opinion were never strongly correlated with major events of the opposition movement or some of the more publicized nuclear accidents.

The most remarkable difference between the French and American nuclear power programs is their contrasting histories of support for or opposition to nuclear power. As Jasper (1988) has shown, U.S. polls continued to show a majority in favor of constructing more nuclear power plants up until 1978, when this support began to decline sharply. The views were nearly even by the time of Three Mile Island in March 1979, a balance that held up remarkably until 1981-1982, when favorable views began to fall dramatically. By the time of Chernobyl in April 1986, three out of four Americans were opposed to new plant construction.

In France, support and opposition were more or less balanced from 1974 to 1979, with the opposition gaining a temporary upper hand: sixty percent against nuclear power and forty percent for it during the peak of the anti-nuclear protests of 1977-1978.[13] At the time of Three Mile Island, public opinion slightly favored the EdF expansion program by the same margins, a superior position that essentially held for several years and then began to increase slowly. By 1986, continued nuclear power growth was favored by about 70 percent of the French public, in remarkable contrast to the U.S.'s experience.

In Sweden, the balance between pro- and anti-nuclear viewpoints was continuous throughout the entire period 1973-1987. Although the anti-nuclear view had the upper hand in the protest- and spillover-filled mid-1970s and again in 1986, there was in fact a slight majority favoring new plant construction during the five years after Three Mile Island. The results of the 1980 national referendum, however, foreclosed this possibility.

Although there are no comparable polls for Germany, the highly visible portion of the anti-nuclear movement peaked with the clashes at Grohnde in 1977,[14] after which it abandoned confrontation as ineffective and potentially alienating. In contrast, more broadly based opposition grew slowly during the 1980s. As with the U.S., the major turning point probably occurred during the early 1980s, well after Three Mile Island but before Chernobyl.[15]

These responses present a persuasive argument against the notion that such factors as nuclear power plant safety or radiological risk dominated the public or political response, regardless of how much they dominated the reporting of the debate. This conclusion supports the argument that what was at stake was not merely nuclear power as a technology but the organizational form of modern industrial society. Comparative work done at the time (and since) points to the importance of more general social movements away from general acceptance of technocratic decision-making and hierarchical organizational forms.

I argue that the fundamental problem of the nuclear power system was its inability to adapt to changing social demands. Only where belief in technocracy and acceptance of central authority and large, hierarchical organizations remained high — or was enforced, as in the former Soviet Union — did nuclear power continue to be deployed in its original form.

Part Five:
Conclusion

Balogh has characterized the nuclear enterprise in the United States as representative of what he calls the "prominstrative" state (professional and administrative).[16] In his view, the development of nuclear power shows how professionals, scientists and engineers became administrators to promote a huge scientific-technical-industrial enterprise. This describes France, Germany and the U.S., particularly in the critical developmental years of the 1960s. Where they differed was in the degree to which the centralized, "prominstrative" technocratic system was open to public participation; in other words, the outcome depended upon the ability or inability of the technocratic elite to keep the system closed off from more general social and political influences. No other indicator separates as clearly the cases of France and the United States.

The factors that contributed to success in full deployment in France were:

1. a willingness to defer to formally designated technical and scientific cadres and a historical tendency to emphasize centralized, hierarchical and non-participatory modalities of political and social control,
2. a relatively closed and powerful scientific-technical elite, which made it difficult to find and organize expert testimony that opposed official government positions,
3. a tendency to encourage formalization and standardization from central authorities both for technological development and for social behavior,
4. a lack of any socially and politically acceptable route to independently criticize or participate in technical and industrial affairs in the face of governmental planning and exertion of central authority,[17]

5. an extremely corporatist readiness to accept standardization of suppliers and equipment rather than encourage diversity and competition. This not only supported the existing hierarchy but also encouraged a rapid movement to scale with existing technologies rather than a search for perfection through continuous innovation,

6. the "EdF factor," namely a single integrated electrical grid under parallel central control, run by a national authority and co-designed to guarantee the smooth integration of large blocks of new generation capacity.

The French successfully deployed nuclear electrification including the grid as a rationalized, standardized technical system of a large scale, without raising much general concern over the potential social price — indeed, without engendering much critical discussion at all. This is a fully realized representation of what the traditional literature characterizes as "modernization," a social trend that even Jacques Tati's legendarily frustrated Mr. Hulot understood to be problematic even as it was accepted as inevitable.

Crozier has described the centralized French system as facilitating control by informally organized technical elites.[18] He further asserts that other characteristics of French life lead to acceptance of considerable restriction of activity by bureaucracies and other formal organizations. Thus, centralization and bureaucratic control are deeply embedded in and reinforced by the national culture and social institutions, complemented by a traditional resistance to opening up the inherently "proministrative" processes to public debate or participation even at the local level.

The apogee of the CEA-EdF nuclear program took place just when the historical tendency towards centralization inherited from the *ancien régime* was beginning to break down. The election of Mitterand in 1982 marked the beginning of a transition in French society that is slowly transferring power out of the hands of Paris bureaucracies, including the most technical and heretofore untouchables of the grands corps. Thus, I re-emphasize that my conclusions apply to the system as it reached its political and technical apogee in the late 1970s and early 1980s. At that time new power plants were being built everywhere, the Super-Phenix fast-breeder was reaching completion, the plutonium facilities at La Hague and the waste plant at Marcoule were being expanded, and Framatome and Cogema were searching for overseas contracts for fuel, fuel services, reactors and other equipment.

American social and political tendencies in the 1950s and early 1960s were much closer to France's than is generally the case.[19] Such phenomena as McCarthyism, the general tendency towards social conformity, the lack of effective oppositional centers even within the universities, and the emphasis on the central importance of scientific and technical careers all point to a society that was evolving in a generally technocratic direction. It was the era of large-scale scientific projects — military and civil — and of promotion of electrical networking and the all-electric home. But the essential resistance of Americans to standardization and central control, particularly in matters of technology, began to reassert itself profoundly in the 1960s.

Analysts of American politics are still studying the origins of the broad political movement whose basis is criticism of and opposition to any and all governmental entities. Whatever the reasoning, there is general agreement that by the end of the 1960s the U.S. — in very strong and dramatic contrast to France — was no longer a society where technocratic elites and techno-industrial bureaucracies (or other remote, hierarchical, and centralized actors) could create and deploy a large-scale technical system *from the top down* without considerable public scrutiny and in many cases coordinated and effective public criticism and intervention.[20]

The situation in Germany posed yet a third case. German technocracy was in some ways halfway between those of the U.S. and France. In its origins, the nuclear program was centralized, comprehensive and state administered. Although the implementation was primarily through the private sector, the tightly-coupled corporate structure of Germany was such that it was sometimes difficult to distinguish what constituted public policy. The peak of German support for nuclear power occurred in the early 1970s, well after the crash of the great bandwagon in America. While the French suppression of protest and refusal to share power effectively killed the anti-nuclear movement in the late 1970s, some parts of the German movement moved away from confrontation as they began to find access to real political power (at first through the Länder) and their own basis of scientific and technical support.

As in the U.S., concern over social externalities became a vehicle for preventing completion of the fuel cycle via plutonium use and the fast-breeder reactor. Germany lacked the central mechanisms of planning and control, a national power grid and a national utility to subsidize the system or provide guaranteed markets. Even the centralization and standardization of reactor production and design through the Siemens-KWU consortium was not enough to prevent the system from reconfiguring into a series of isolated power plants stripped of systemic context. By the 1990s, Germany, which had started out to emulate France, had reconfigured its nuclear system so it looked more like that of the United States.

Finally, Sweden was also a very technocratic state and made a serious attempt to build a power system without a national fuel cycle: it externalized the plutonium part but retained its own waste disposal program. Sweden made heavy investments in nuclear power in the late 1960s, primarily on the basis of central planning and control along a European corporate model similar to that of Germany. Sweden also had considerable public tolerance for central planning and control, particularly in technical matters. At the same time, Sweden was even more open than Germany to the emerging anti-technocratic social trends. Moreover, its dissidents had greater access to counter-expertise, and the consensus-seeking nature of Swedish politics made the government even more sensitive to organized protest than in the U.S. Unable to isolate its opponents and shield its techno-industrial bureaucrats from politics, Sweden actually had a non-Socialist coalition running — and winning — on an anti-nuclear platform against the Socialists in 1976. If the 1980 referendum to completely phase out nuclear power is adhered to, it will be unique. But even if nuclear power is retained, it will again be in a reconfigured form similar to that of Germany, and to a lesser extent, the U.S.

Table 1 (see pp. 254-255) presents a cross-national comparison of some of the factors and data for the four countries during the critical decade of the 1970s.

In France, there was a highly technocratic society, centralization, national government control and a power grid that grew in a centralized pattern. Until well into the 1980s, the socio-technical pattern of systems accretion emphasized increasing integration of the grid and the nuclear power system into a single, coordinated overall system for the delivery of electricity. In the U.S. and Germany, the trends were towards less hierarchy and less centralization, but with a growing tendency to superimpose democratic and participatory decision-making in areas once ruled by technocratic and exclusive ones. The pattern of decentralized and idiosyncratic electrical utilities provided fertile ground for adaptation to the emerging socio-political trends in the design and implementation of the grid. Sweden was somewhat in between. Although Sweden retained some of the overall integrative elements of the French case, by the early 1970s it became increasingly difficult to impose a planned system upon a society in which participation in decision-making was high and technocratic rule was questioned.

In sum, the nuclear power system in these four countries was in its inception and implementation socio-technically inflexible. It placed demands on society for conformity and long-term stability without itself having much room for reciprocal adaptation. As both electrical network and society moved towards increasing socio-technical flexibility, the nuclear system found itself unable to adapt gracefully. Instead, the system underwent a radical reconfiguration which first divided the power plants from the fuel cycle and then separated nuclear programs from overall electrical planning.

Only in France did the original conditions by which the nuclear industry developed remain stable enough over the critical deployment period to allow for the implementation of a full nuclear fuel cycle and the complete integration of a national distribution and marketing system.

Even in France, however, the situation has grown more complex and less predictable. The French success with the nuclear power system depended upon social and political factors and organizational and institutional arrangements that have profoundly changed since the 1970s when the deployment decisions were made. The scientific-technical elite are still strong but they no longer rule — even in this area. The government and other decision makers may be difficult to influence compared to those in Germany or the U.S., but they are far more open than before. Centralization and rigidity in French planning has greatly diminished and the French public is less willing to accept the dictates of centralized bodies, even those as powerful and familiar as EdF. If that is the case, the present difficulties in the French nuclear program are deep and structural and not simply the transient effects of decreased growth in energy consumption and the current economic recession.

Epilogue

The classic nuclear fuel cycle represents a "techno-modern" socio-political form whose features include the following: considerable autonomous power for scientists and technicians, centralized authority and policy-making, long-term inflexible deterministic planning, large, complex and highly specialized plants and processes tightly coupled into a technically rigid system, and a requirement for hierarchical, administrative and managerial practices to guarantee the required stability both during deployment and when fully deployed.

TABLE 1 A Four-Country Comparison

	U.S.	France	Germany	Sweden
Scientific-technical development	Government-private mix	Central governmental	Central corporatist	Central corporatist
Scientific and technical elites	Open, extended, advisory	Closed, empowered	Semi-closed, empowered	Open, empowered
Reactor manufacture and sale	Private, competitive	Central, quasi-governmental	Mixed, private-corporatist	Corporatist (joint govt.-private)
Reactor standardization	Very little, company driven	Complete, centralized	De facto	Mixed
Nuclear fuel cycle supporting services	Mixed governmental and private; not assured	Governmental: complete, centralized, and assured	Mixed central and corporatist; assured but not deployed	Exernalized (except for wastes)
Electrical utility structure	Private, decentralized, fragmented	Public, highly centralized, and integrated	Corporatist-private, partially centralized but fragmented	Mixed national and private; integrated.
Electrical system planning	Fully private; little governmental input	Public, national, and fully central	Regionalized with some central input	Mixed national and private, fragmented
Electrical grid (and Market)	Fragmented, with regional coordination. (Negotiated)	Nationally integrated and operated (Guaranteed)	Regional-fragmented with coordination. (Semi-planned)	Partially integrated (Semi-planned)

Political milieu	Coordinating and encouraging	Controlling and planned	Coordinating and facilitating	Mix of planned and facilitating
Political environment	Multiple, overlapping, fragmented	Central, national, integrated	Mixed regional and national	Centralized plus heavy public input
Policy formulation[a]	Centralized, fragmented, closed	Centralized, unified, closed	Centralizeded, unified, closed	Centralized, unified, open
Policy[a] implementation	Decentralized, fragmented, open	Centralized, unified, closed	Decentralized, fragmented, open	Centralized, unified, closed
Social environment for hierarchical technical systems	Challenging, independent, and techno-suspicious	Accepting conformal, and fully technocratic	Accepting conformal, industrial technocratic	Mixed conformal, industrial technocratic
Character of original dissent	Largely in-system, becoming fully in	Minority oppositional, excluded	Minority alienated, moved in-system	Broad, completely in-system
Elite participation	High	Almost nil	Very low	Moderate to high
Methods for resolving socio-technical conflict	Negotiation and cooptation, leading to power sharing	Exclusion, rejection, denial of legitimacy	Exclusion, then negotiation leading to sharing of power	Negotiation and search for social consensus

a From Campbell, 1988.

Nuclear power is inherently and structurally techno-modern. If advanced industrial societies continue to move, as they have over the past decade, towards more flexibility, adaptability and participation with shorter decision and implementation times and looser and less hierarchical coupling, then there is very little chance that nuclear power can be revived. The U.S. and German systems will continue on their present course until the plants age and shut down. Even the French system will gradually lose legitimacy, stirring a lively debate about the possibilities for a post-nuclear electric grid.

Present debates about improved reactor designs may simply be epiphenomenal. The issue is not whether the plants can be made smaller, safer or more cost-effective but whether the system of which they will be a part will be acceptable. As long as widespread use of nuclear power continues to require deployment of a complex, large-scale and rigid system, there is no adaptation, evolution, or modification that will reduce the political and social conflict that surrounds it.

There is a more somber possibility. In a more heavily populated and technological world, loss of autonomy and increased technical, industrial and inter-societal (as well as inter-personal) coupling seem nearly inevitable. There are already some signs of strain from deregulation in industries within such areas as transportation, banking and telecommunications. The current wave of quasi-democratic decentralization and deconcentration may, therefore, prove to be transient or illusory — or simply fail.

What this implies for nuclear power's future will depend upon the state of the technology and industry, at some future point of transition in an environment that may be radically different. Meanwhile, the present decline in the growth and legitimation of nuclear technology seriously threatens any possibility for further progress in either reactor or system design.

If the current trends are long term, there is every reason to believe that the requirement for comprehensive, advance systems-level planning is increasingly incompatible with the current state of social, cultural and political development in most of the advanced industrial countries. The re-birth of nuclear power will then have to await yet another transition to a post-postmodern form in which the banners of positive planning, centralization, hierarchy and technocracy once again fly over the capitals of Europe and America.

Acknowledgements

I wish to thank Mathilde Bourrier, Gabrielle Hecht and Arne Kaijser for their advice and help in summarizing the French and Swedish experiences, as well as to apologize to the many authors I drew upon for not citing them in detail in this brief review. Any omissions or errors that may remain are my own.

Notes

1. The plutonium reactors were to be harnessed to the electrical grid, but U.S. law kept even that military-commercial link problematic throughout its history.

2. This was the CANDU reactor, a natural-uranium fueled, heavy-water moderated and light-water cooled reactor. Successors to that original conception are still being planned. Considerable electrical power is exported to the northeast U.S. from power generated from CANDUS.

3. Lowen, 1987.

4. Although several U.S. industrial firms offered in 1951-1952 to build civil reactors if the government would guarantee to purchase weapons material as a form of subsidy, the AEC decided otherwise in late 1952, at least partially because its own facilities seemed adequate.

5. Allen, 1977.

6. Rochlin, 1979.

7. Rochlin, 1979.

8. Hirsh, 1990.

9. The full name was the Generic Environmental Impact Statement on the Use of Mixed Oxide (i.e. mixed uranium-oxide and plutonium-oxide) Fuel in the Light-Water Reactor Fuel Cycle.

10. Although the experimental *Rapsodie* fast-breeder was already under way, it was 1962 before CEA began talking publicly about civilian use of plutonium. The subsequent tension between CEA and EdF over the use of civil plutonium for military purposes and plutonium credits is well covered in Hecht, 1992.

11. Kaijser, 1992.

12. Notable exceptions are Balogh (1991) and Thomas (1988). Also interesting are the social movement oriented analyses of Jasper (1988) and Joppke (1989).

13. Unfortunately, it is nearly impossible to correlate public opinion polling cross-nationally. According to Nelkin and Pollak (1980), public opinion was favorable to nuclear energy in 1974 both in France (74 percent in favor, 17 percent opposed) and in Germany (60 percent in favor, 16 percent opposed). Although support had sharply declined in both countries by 1978 (47:42 in France, 53:43 in Germany) overall opinion remained slighly positive. Note that

public support had dropped sharply in both countries even before Three Mile Island.

14. Joppke, 1989.

15. Both major parties were strongly pro-nuclear during 1979 when the debate over the reprocessing plant at Gorleben was at its peak.

16. Balogh, 1991.

17. Any truly oppositional movement had to take the alternate route of confrontation and demonstration for lack of a voice within the cadres of the grands corps.

18. Crozier, 1964.

19. Balogh, 1991.

20. The emphasis is meant to distinguish other types of large-scale system penetration, such as telefaxes and personal computers that were *perceived* by the public to be non-central, participatory, and democratic.

References

Albonetti, Achille. 1972. "Europe and Nuclear Energy." Paris: Atlantic Institute for International Affairs. *Atlantic Papers* no. 2.

Allen, Wendy. 1977. "Nuclear Reactors for Generating Electricity: U.S. Development from 1946 to 1963." Santa Monica, Ca.: Rand Corp. Report R-2116-NSF.

Balogh, Brian. 1991. *Chain Reaction: Expert Debate and Public Participation in American Commercial Nuclear Power, 1945-75*. Cambridge: Cambridge University Press.

Bupp, Irvin C. and Jean-Claude Derian. 1978. *Light Water: How the Nuclear Dream Dissolved*. New York: Basic Books.

Campbell, John N. 1988. *Collapse of an Industry: Nuclear Power and the Contradictions of U.S. Policy*. Ithaca: Cornell University Press.

Cantelon, Philip L., Richard G. Hewlett, and Robert C. Williams, eds. 1992. *The American Atom: A Documentary History of Nuclear Policies from the Discovery of Fission to the Present* (Second Ed.). Philadelphia: University of Pennsylvania Press.

Cardot, Fabienne, ed. 1986. *1880-1980: Un Siècle d'Électricité Dans Le Monde*. Histoire de l'Électricité, ed. F. Cardot. Paris: Presses Universitaires de France.

_____ 1988. *Des Enterprises pour Produire de l'Électricité*. Histoire de l'Électricité, ed. F. Cardot. Paris: Presses Universitaires de France.

Cochran, Thomas B. 1974. *The Liquid Metal Fast Breeder Reactor*. Washington D.C.: Resources for the Future.

Davis, Mary D. 1988. *The Military-Civilian Link: A Guide to the French Nuclear Industry*. Boulder, Co.: Westview Press.

Dawson, Frank G. 1976. *Nuclear Power: Development and Management of a Technology*. Seattle: University of Washington Press.

Dorget, François. 1984. *Le Choix Nucleaire Français*. Paris: Economica.

Dupuy, François and Jean-Claude Thoenig. 1983. *Sociologie de l'Administration Française*. Paris: Armand Colin.

Forsberg, Charles W. and William J. Reich. 1991. "Worldwide Advanced Nuclear Power Reactors with Passive and Inherent Safety: What, Why, How, and Who." Oak Ridge Tn: Oak Ridge National Laboratory. Report ORNL/TM-11907 (September).

Galtung, Johan. 1981. "Structure, Culture, and Intellectual Style: An Essay Comparing Saxonic, Teutonic, Gallic, and Nipponic Approaches." *International Review of Administrative Sciences*, vol. 20 no. 6: 817-856.

Goldschmidt, Bertrand. 1964. *The Atomic Adventure*. Oxford: Pergamon Press.

_____. 1967. *Les Rivalités Atomiques*. Paris: Librairie Arthème Fayard.

_____. 1982. *The Atomic Complex: A Worldwide Political History of Nuclear Energy*. Bruce M. Adkins, trans. La Grange Park, IL.: American Nuclear Society.

Harrison, Michael M. 1981. *The Reluctant Ally*. Baltimore: The Johns Hopkins University Press.

Hecht, Gabrielle. 1992. *The Reactor in the Vineyard: Technological Choice and Cultural Change in the French Nuclear Program, 1945-1969*. Ph.D. Dissertation, University of Pennsylvania.

Hewlett, Richard D. and Oscar E. Anderson Jr. 1962. *The New World, 1939/1946*. A History of the United States Atomic Energy Commission, vol. I. University Park, Pa.: The Pennsylvania State University Press.

Hirsh, Richard F. 1989. *Technology and Transformation in the American Electric Utility Industry*. Cambridge: Cambridge University Press.

Hughes, Thomas P. 1983. *Networks of Power: Electrification in Western Society 1880-1930*. Baltimore: The Johns Hopkins University Press.

_____. 1986. "Visions of Electrification and Social Change." In *1880-1980: Un Siècle d'Électricité dans le Monde*, edited by F. Cardot, 328-340. Paris: Presses Universitaires de France.

_____. 1987. "The Evolution of Large Technical Systems." In *The Social Construction of Technological Systems*, edited by W. E. Bjiker and T. P. Hughes, 51-82. Cambridge, Ma: MIT Press.

Jasper, James. 1990. *Nuclear Politics: Energy and the State in the United States, Sweden, and France*. Princeton: Princeton University Press.

Joppke, Christian Georg-Maximilian. 1989. "The Social Struggle Over Nuclear Energy: A Comparison of West Germany and the United States of America." Ph.D. diss., University of California, Berkeley.

Kaijser, Arne. 1986. "From Local Networks to National Systems." In *1880-1980: Un Siècle d'Électricité dans le Monde*, edited by F. Cardot, 7-22. Paris: Presses Universitaires de France.

_____. 1992. "Redirecting Power: Swedish Nuclear Power Policies in Historical Perspective." In *Annual Reviews of Energy and the Environment*, vol. 17, 437-462. New York: Annual Reviews, Inc.

Keating, Michael and Paul Hainsworth. 1986. *Decentralisation and Change in Contemporary France*. Aldershot, Hants: Gower Publishing Company, Ltd.

Kern, Horst and Michael Schumann. 1989. "New Concepts of Production in West German Plants." In *Industry and Politics in West Germany*, edited by P. J. Katzenstein, 87-111. Ithaca: Cornell University Press.

Kohl, Wilfrid L. 1971. *French Nuclear Diplomacy*. Princeton: Princeton University Press.

Lilienthal, David. 1963. *Change, Hope, and the Bomb*. Princeton: Princeton University Press.

_____. 1980. *Atomic Energy: A New Start*. New York: Harper & Row.

Lönnroth, Måns, Peter Steen, and Thomas B. Johansson. 1980. *Energy in Transition: A Report on Energy Policy and Future Options*. Rudy Feichtner, trans. Berkeley and Los Angeles: University of California Press.

Lowen, Rebecca S. 1987. "Entering the Atomic Power Race: Science, Industry, and Government." *Political Science Quarterly*, vol. 102 no. 3 (Fall): 459-480.

Löwgren, Marianne. 1992. "Nuclear Waste Management in Sweden: Balancing Risk Perceptions and Developing Community Consensus." In *88th Annual Meeting of the American Political Science Association*, September 3-6, Chicago:

Lucas, N. J. D. 1979. *Energy in France: Planning, Politics and Policy*. London: Europa Publications Ltd.

Massey, Andrew. 1988. *Technocrats and Nuclear Politics*. Aldershot, Hants: Gower Publishing Compant, Ltd.

Morone, Joseph G. and Edward J. Woodhouse. 1989. *The Demise of Nuclear Energy?: Lessons for Democratic Control of Technology*. New Haven, Ct: Yale University Press.

Müller, Wolfgang D. 1990. *Geschichte der Kernerenergie in der Bundesrepublik Deutschland*. Stuttgart: Schäffer Verlag für Wirtschaft und Steuern GmbH.

Nau, Henry R. 1974. *National Politics and International Technology: Nuclear Reactor Development in Western Europe*. Baltimore: The Johns Hopkins University Press.

Nealey, Stanley M., Barbara D. Melber, and William L. Rankin. 1983. *Public Opinion and Nuclear Energy*. Lexington, Ma.: Lexington Books.

Nelkin, Dorothy and Michael Pollak. 1982. *The Atom Besieged: Antinuclear Movements in France and Germany*. Cambridge, Ma: The MIT Press.

Nichols, Elizabeth and Aaron Wildavsky. 1987. "Nuclear Power Regulation: Seeking Safety, Doing Harm?" *Regulation*, vol. 11 no. 1: 45-53.

Noiret, René. 1988. "La Commission des marchés d'Électricité de France." In *Des Enterprises pour Produire de l'Électricité: Le Génie Civil, la Construction Electrique, les installateurs*, edited by F. Cardot, 361-380. Paris: Presses Universitaires de France.

Pasqualetti, Martin J. and K. David Pijawka, eds. 1984. *Nuclear Power: Assessing and Managing Hazardous Technology*. Boulder, Co.: Westview Press.

Pringle, Peter and James Spigelman. 1981. *The Nuclear Barons*. New York: Holt, Rinehart and Winston.

Rochlin, Gene I. 1979. *Plutonium, Power, and Politics: International Arrangements for the Disposition of Spent Nuclear Fuel*. Berkeley and Los Angeles: University of California Press.

Scheinman, Lawrence. 1965. *Atomic Energy Policy in France Under the Fourth Republic*. Princeton: Princeton University Press.

Schoenbaum, Thomas J. and Joseph H. Ainley. 1988. *The Regulation of Nuclear Power in Three Countries: The United States, France, and Japan*. A Dean Rusk Center Monograph. Athens, Ga.: The Dean Rusk Center for International and Comparative Law, University of Georgia School of Law.

Thoenig, Jean-Claude. 1987. *L'Ère des Technocrates: Le Cas des Ponts et Chaussées*. Paris: Editions l'Harmattan.

Thomas, Steve D. 1988. *The Realities of Nuclear Power: International Economic and Regulatory Experience*. Cambridge: Cambridge University Press.

Touraine, Alain, Zsuzsa Hegedus, François Dubet, and Michel Wieviorka. 1983. *Anti-Nuclear Protest: The Opposition to Nuclear Energy in France*. Peter Fawcett, trans. Cambridge: Cambridge University Press.

U.S. Congress, Office of Technology Assessment. 1984. "Nuclear Power in an Age of Uncertainty." Washington, D.C.: U.S. GPO. OTA-E-216 (February).

Van Buiren, Shirley. 1980. *Die Kernenergie-Kontroverse im Spiegel der Tageszeitungen*. Sozialwissenschaftliche Reihe des Battelle-Instituts e.V., vol. 1, edited by J. Scharioth and R. V. Gizycki. Munich: R. Oldenbourg Verlag.

PART FOUR

Controlling Car Traffic: Will the System Change?

12

Car Traffic at the Crossroads: New Technologies for Cars, Traffic Systems, and Their Interlocking

Reiner Grundmann

All the misfortunes of men
derive from one single thing,
which is their inability
to be at ease in a room (at home).

(Pascal)

Introduction

This chapter examines the existing literature on large technical systems (LTS) and tries to apply some findings to the development of the car traffic system. I will first present some theoretical elaborations from the LTS discourse which offer definitions, properties and dynamics of large technical systems. Then I will apply them to the car traffic system and elaborate on how this system will undergo a radical reconfiguration. I will discuss three cases of empirical material, namely different systems of motive power, transit systems and attempts at automation via electronic road management. Throughout the chapter, several theoretical models are tested, including the system builder approach of Thomas Hughes, the interorganizational network approach of Bernward Joerges and the actor-network approach of Michel Callon and Bruno Latour.

Part One:
A Large Conceptual System

"The American Highway network is the most costly public-networks construction effort in human history and the most important visible symbol of the transformation of the American landscape by automobility."[1] A similar judgment would hold true for most western countries. However, until now car traffic networks have not been analyzed by scholars working in the field of Large Technical Systems. It seems time to correct this situation.

If one looks at the literature on technical systems, different approaches, definitions and fields of study will be detected. In Hughes' model, technical systems "embody the physical, intellectual, and symbolic resources of the society that constructs them."[2] Hughes stresses the need for control of these systems and for them to be directed towards goals. "The goal of an electric production system, for example, is to transform available energy supply, or input, into desired output, or demand."[3]

In *Networks of Power*, Hughes describes the diminishing of technical alternatives, the strengthening of networks, and the momentum of systems: "These sociotechnical systems had high momentum, force and direction because of their institutionally structured nature, heavy capital investments, supportive legislation, and the commitment of know-how and experience. This momentum was a conservative force reacting against abrupt changes in the line of development. Because of the conservative momentum, rarely were radical inventions, technical or social, introduced."[4]

Another important feature of technical systems, according to Hughes, is the tendency toward maximum coverage of the net. "In modern industrial nations technological systems tend to expand, as shown by electric, telephone, radio, weapon, automobile production, and other systems. A major explanation for this growth, and one rarely stressed by technical, economic, or business historians, is the drive for high diversity and load factors and a good economic mix...The load factor is the ratio of average output to the maximum output during a specified period."[5]

However, LTS are also unplanned and often show unintended effects, as Joerges notes:

> Retrospective studies of LTS show that they never develop according to the designs and projections of dominant actors: LTS evolve behind the backs of the system builders, as it were. It has been shown... by LaPorte... that typically none of the agencies contained in LTS manage to form a somewhat complete picture of their workings. LTS seem to surpass the

capacity for reflexive action of actors responsible for operating, regulating, managing and redesigning them in ways which, as social scientists, we understand poorly.[6]

It might be a rather scholastic exercise to give a watertight definition of a Large Technical System. However, as Joerges points out:

Some types of technical systems can be singled out as undisputably large: those complex and heterogeneous systems of physical structures and complex machineries which (1) are materially integrated, or "coupled" over large spans of space and time, quite irrespective of their particular cultural, political, economic and corporate make-up, and (2) support or sustain the functioning of very large numbers of other technical systems, whose organizations they thereby link.[7]

An LTS, therefore, is not a single manufacturing enterprise or firm.[8] As Joerges makes clear, "LTS are not technical systems contained in, or co-extensive with identifiable organizations and their external reaches. Rather, LTS contain many organizations. Some of these wholly merge with LTS, others only partially, some are concerned with operating their technical subsystems, some with other, non-technical components of LTS."[9]

However, LTS also represent a societal dilemma. "They are hard to black box for good, have an irreducible potential for controversy, and their integrations with their social base remain precarious because for structural reasons strategies aiming at closure tend to reproduce conflict on a larger scale."[10]

The process of black boxing (i.e. problem solving) plays a prominent role in the work of Latour and Callon. During the innovative process, a network of heterogenous elements emerges which must become stable if the innovation is to succeed. As Latour puts it, a black box can be a scientific text or a machine, "a well-established claim" or an "unproblematic object."[11] This means that humans and non-humans act together in the construction of a new technology. At the same time as the new technology emerges, a support network comes into being. Both support each other and no determinism is implied. At first sight, this approach has no direct relationship to the problem of large technical systems. However, I will suggest that it can be used — as a radical extension of Hughes' and an illumination of Joerges' approaches — to present some of the underlying problems in an interesting way.

Is Car Traffic an LTS?

Does the car traffic system belong to the family of LTS? Let us take, for example, the criterion that the elements of the system have to be coupled in order to speak of a system. Then all depends on the definition of the components of the system. If we take roads, traffic signs, rules and norms of regulation, cars, and trucks as components of the system, it would not be disputed how tightly coupled these components are. On the other hand, if we define automobile firms, gas and service stations, roads, bridges, tunnels, local authorities, hospitals, cars, drivers, and interest groups as the components of the system, the question of whether the system is tightly or loosely coupled might not get a simple answer. Renate Mayntz writes that it may be difficult for an uncritical observer to detect the system properties in the example of the automobile: "The coupling of different components of the automobile system ... is so loose that a passing observer might not even see a system."[12]

Another feature of technical systems is the load factor mentioned earlier, which is important for the growth of many technical systems and their integration. The operational temptation to achieve maximum coverage is certainly present in most traffic systems.[13] In the projects discussed here, the intention is to increase the load factor within single systems and through tighter coupling of different traffic systems. It is estimated that a load increase of thirty percent is possible through more efficient traffic management.[14]

However, the maximum load is not always the optimum load. Let us take the example of highways. Here the maximum load is achieved at an average speed of 16.2 km/h. At this speed, 111 cars can be accommodated on each kilometer compared to ten cars at the speed of 200 km/h (in Germany there is no speed limit on highways). However, this maximum coverage would hardly be regarded as travelling. A load factor of forty-eight percent per lane would still permit a top speed of 100 km/h.[15]

Part Two:
The Future of the Car Traffic System

In this chapter, I will focus on the following questions. In what sense will the car traffic system undergo a radical reconfiguration and what are the dynamics of this process? Is there a pattern of technical development? Can the innovative technology be found in niches, and will different technologies co-exist or be substituted?

These questions will be explored using three examples, namely the possible revival of the electric vehicle, the growing importance of public transport, and electronic road management.

Towards a Revival of the Electric Vehicle?

Growing environmental awareness and tighter emission standards have forced the automobile industry to develop new technical solutions to the problem of pollution and depletion of natural resources. Since the 1970s, programs have been underway to reduce gasoline consumption and exhaust fume emissions, including the design of new engines and the introduction of the lean-burn engine.

The general public, as well as parts of the political and legal system, have long felt that more radical measures were necessary. The emerging unrest in the automobile sector and a perceivable discontent with traditional car and motor systems has reached the top firms in this industry.

However, what the car industry has to offer is far from envisaged environmental norms. As several studies have shown, road traffic plays a paramount role in pollution. Although there has been a reduction of pollutants (except particles), the emissions of CO_2 have increased and will further increase. Road traffic produces 20 percent of all CO_2 emissions, of which automobiles alone make up 75 percent.[16] The electric engine seems to perform better with regard to NO_x, CO and HC emissions and worse with regard to SO_2 emissions because German (mainly coal-based) electric power plants deliver thirty percent of electricity. The problems with electric vehicles can only be avoided if electricity can be generated through non-fossil fuels. It is not surprising that the nuclear energy lobby has spotted new opportunities to praise its services. The emissions of CO_2 will also increase, according to an estimate of the German environmental agency *Bundesumweltamt*. An electric vehicle will produce 24 kilograms of CO_2 every 100 km compared to 20.5 kilo with the internal combustion engine.

Black Boxing: Too Many Unresolved Problems and Variables in Flux

It is interesting that the elements of a new generation of cars have existed for a long time.[17] Thus, it is astonishing that few of these elements have been put into practice until now, says Frederic Vester, author of *Ausfahrt Zukunft*, a study commissioned by Ford Cologne some years ago. "There are light cars, short cars, high cars, comfortable cars, solar cars, cars which run without gasoline, cars with Stirling engines, with high-tech-electronics ... But all are far away from what a systems logic would demand."[18] Vester blames this shortcoming on the linear thinking of the responsible managers, who only focus on a single problem at a time. But Vester's lesson is "if you change only one component at a time, you will have no success." If it does not want to suffer the same fate as dinosaurs, the car industry will have to learn to think in system terms and become courageous enough to change many components simultaneously.

Vester's illuminating account contrasts with findings in current literature in the history and sociology of technology that stress that in order to be successful, engineers, business managers and system builders solve *critical problems* or dream up entire new sociotechnical worlds. This process of problem-solving is also known as "black boxing." Usually these system builders identify a range of problems which must be solved if the technology is to be stabilized. In the case of the ecological car, many variables need to be solved simultaneously on a purely technical level. Vester is right that a radical change in the car's components will also require a change in the overall traffic system because the electric cars he envisioned are slower, lighter and shorter than traditional cars. He is left with the "chicken and egg" problem: an electric car will not be developed without adequate demand and infrastructure, while the infrastructure will not be developed without adequate demand and new vehicles. Such demand is unlikely to materialize in the absence of adequate infrastrucure and electric vehicles at competitive prices. And unfortunately, a system builder who could unify all the recalcitrant elements is not in sight. It seems as if there are so many problems (Hughes' reverse salients) that a potential system builder would not even know where to start.[19]

Creating Networks

At this point, the proposed methodology by Latour and others can help us.[20] They emphasize creating networks as an important point in the process of establishing facts and artifacts. With the electric vehicle, this network takes shape in the form of an alliance between electricity firms (including power plants), battery devices, car designers, environmentally conscious consumers, political leaders, local authorities, and interest groups. Callon and Latour's quasi-Machiavellian approach emphasizes that alliances have to give the same weight to artifacts as to human actors. If the artifacts cannot be stabilized, the whole network falls apart. It is intriguing that Callon and Latour have also studied the fate of the electric car.[21] In the battle between Renault and the electric industry in France some twenty years ago, Renault emerged as the clear winner. And, *pace Vester*, Renault's opponent EDF did in fact try to bring together many elements into a new system.

In the German case some years later, we have different actors and artifacts (different 'actants' in Callon and Latour's terms) with the most important being the entry of a new element, namely CO_2. Since the public debate focuses on global warming and other environmental problems related to CO_2 emissions, the electric car does not solve as many problems as was hoped.

The time dimension also becomes crucial here. It is *early* problem solving which matters. One hundred years' experience with the internal combustion engine and twenty years' experience with gasoline-saving technologies have given the dominant design a solid advantage. If the engineers develop an internal combustion engine that consumes around three liters every 100 km,[22] the energy balance could be similar to that of electric cars. By improving catalytic-converter technology, the dominant design would be given another chance to survive. To be sure, such a technical fix would not be sufficient for the forthcoming Californian legislation which envisages four progressively stricter categories of vehicles (transitional-low emission vehicles [TLEVs], low-emission vehicles [LEVs], ultra low-emission vehicles [ULEVs] and zero-emission vehicles [ZEVs]).[23] However, if at the end of the century, the composition of the car fleet changes towards ULEVs, legislation might again become more lenient. Up until now, it has not been clear if the Californian project will also be dominant in Europe.

Conservative Innovations

Over twenty years ago, Lawrence White remarked:

Perhaps the most striking thing about automotive technology in the postwar period has been the lack of fundamental change or advance. Cars built in 1968 are not fundamentally different from cars built in 1946... Consequently, in describing the behavior of the auto companies with respect to technical progress, we are dealing with refinements of the technology, innovational advances to bring basic inventions to a marketable state, rather than with the basic inventions themselves.[24]

Rather than to small, ecological cars, the long-run trend has been to larger cars with more power and more luxury options.[25]

Mikael Hård has introduced a helpful distinction between two different types of innovative activity among engineers. He claims that the degree of global regularity is substantial only on the level of technical visions and models (*Leitbilder*), not on the level of local technical solutions. In other words, standard solutions rarely emerge on the level of particular technical solutions but rather on the level of technical goals, archetypes and visions.[26] In the case of the diesel engine, he lists as global patterns the lower engine weight per horsepower, higher speed, more horsepower, and higher torque, especially at lower speeds.[27] If one applies this line of thought to the present case, we would expect one or few global visions and many local solutions. In fact, this is what can be observed. There are some global patterns such as reducing gasoline consumption and emissions, building links between traffic systems, and introducing smart electronic road management. However, local solutions vary considerably.

In Latour's model, the local and global would eventually have to meet again if an innovation is to succeed. "Obligatory passage points" would lead to this local-to-global transformation. Since no such obligatory passage points are visible,[28] we should use the somewhat oxymoronic term "conservative innovations."[29]

For a long time, such conservative innovations consisted of simply substituting the engine with batteries and electric motors. Another example is the device of an automatic stop-start engine. The so-called hybrid vehicle with an internal combustion engine and a small electric motor is yet another example of this conservative approach. These cars embody an engineering approach that tries to build into the existing artifact a device to reduce gasoline consumption and emissions. This could be called a typical "end-of-the-pipe technology." Volkswagen CEO Ferdinand Piëch, aware of the problem, states that the task is not to find a new technical concept but to define a "reduced new vehicle concept."[30]

Thus far, the electrical prototypes from leading car manufacturers have been heavy, expensive and unattractive. This results from the wide use of lead and acid batteries which have modest performance at low costs. However, independent car manufacturers are trying to build *another type of car*, i.e. a smaller and lighter car. These cars weigh around 600 kilos and accelerate slower than cars with internal combustion engines. However, they are able to keep up with speeds in city traffic, operate on batteries that are easily rechargeable, and can be built at competitive prices if they are produced on a large scale.[31]

Several prototypes are now being tested in different cities. Nevertheless, the new technology is still more complementary than substitutive.

This situation will not change soon, although the electric industry is preparing for market shares of electric vehicles. ABB intends to start producing 40,000 batteries annually beginning in 1994. RWE and AEG are preparing to enter the market. The technical characteristics of batteries are different in each company. The batteries' technical problems include safety, recharging times, power, and propulsion systems. As a research team at UCLA put it, there is a trade-off between energy density and useful life. "High energy density provides high power and acceleration as well as long driving ranges. Long battery life is critical for economic reasons. No commerically available battery performs efficiently in both areas; increasing energy density almost inevitably results in reduced battery life."[32] Until now it has not been clear whether electric vehicles will use AC or DC electric motors.

As regards comparative costs, Lenz & Wiescholek maintain that there is no market for the electric vehicle. They calculate a price of 7.1 Pfg./km for an electric-powered Golf as against 6.6 Pfg./km for the conventional model.[33] Their calculation is based on average gasoline prices of 83 Pfg./l (1987); in the meantime these prices have risen again. With regard to the volume of externalities produced by private car traffic, there are only estimates. For Germany, these estimates vary between 45 billion and 100 billion Deutsche Mark. If one would consider this sum for calculating taxes on motor vehicle use, a different pattern of comparative costs would emerge.

The R&D departments of German car manufacturers are mainly concerned with improving the traditional technology of the internal combustion engine, especially by further reducing pollutants by optimizing the engine design and functioning of the catalytic converter. Since the catalytic converter does not prevent CO_2 emissions, further reduction of gasoline consumption and the use of alternative fuels are imperative.

It is interesting to see what argumentation and rhetoric engineers use.[34] Purely technical factors are presented as if they themselves could justify a technical paradigm. Especially important are the notions of efficiency (*Wirkungsgrad*) and energy density (*Energiedichte*) which are traditionally uttered as battle cries by defenders of the internal combustion engine (in Hård's sense, these are global dimensions). Surprisingly enough, the electric industry has also been using this type of argument recently.[35] This hypostatization of technical parameters proves to be fictitious in cases where the parameters of the social world are changing, i.e. when oil prices increase, environmental laws get stricter and consumer needs for cars change. The *Wirkungsgrad* of a motor or the *Energiedichte* of a energy resource is only *one* factor that plays a role in the design of a new engine or a new car concept. To be sure, engineers like their world of *Wirkungsgrad* and they believe that it represents the real core of their design efforts. Consequently, they regard the reduction of emissions as an outside perturbation which forces them to accept compromise.[36]

Turning to the local dimension, German engineers appear to think that the superiority of the catalytic converter in reducing emissions is beyond dispute. The lean burn engine is said to be out of the battle forever.[37] Nevertheless, American, English and Japanese firms have different views on this matter. Toyota, for example, tries to combine the three-way catalyst with the lean burn engine.[38]

As noted above, the same conservatism can be seen in the development of an electric vehicle.[39] Until now, the automobile industry has not been able to get rid of the model of the all-purpose vehicle[40] and has therefore favored retaining the internal combustion engine as the car's technical core. The hybrid car and more fuel-efficient cars are examples. Mainstream thinking in the industry translates the global vision of a less polluting car into local solutions which do not vary much. Therefore, activities within niches become important for radical innovations. It is up to the electric industry and independent designers and engineers (and nations free from automobile corporations such as Switzerland and Austria) to develop a new paradigm of the car.[41] Volkswagen wants to keep in contact with such innovations from niches, as shown by its cooperation with the Swiss company SMH. However, only the future will tell if the "Swatch-Mobile" becomes reality.

Other Ecological Automobiles

There is no reason to be too optimistic about the near future for more ecologically benign automobiles (i.e. including engines based on alternative fuels).

> The potential of alternative fuels to substitute for gasoline varies considerably from country to country and fuel to fuel. In the short run, no single alternative is likely to become a panacea with global applicability. Those that emerge are likely to supplement gasoline, rather than replace it. In the longer run, hydrogen could become a universally used fuel. But an enormous research boost is needed now to make its generation less costly and to achieve breakthroughs in hydrogen vehicle technology.[42]

As noted earlier, the large technical system of automobility was not designed by a single person or firm. Instead, it grew in a globally unplanned way over many decades. Radical changes are inevitable given the system's limits to growth. In this process, the emerging alliances in different networks are decidedly important. The general pattern that defines new visions is global: it involves pollution, congestion, and conservation of resources. The proposed solutions depend on local cultures of e.g. engineering, political alliances, and consumer behavior.

Public Transit Systems and Their Possible Interlocking

The restructuring of the whole traffic system in most European and other cities seems inevitable in the face of the current traffic congestion and pollution. Until recently, however, the owners, managers and users of various means of transportation have considered themselves as part of a "zero sum game" (i.e. I gain what you lose and vice versa). There are some signs that matters are changing and a symbiotic reading will gain importance (i.e. we all profit from each other). The automobile industry has an interest in such interlocking networks since driving becomes more attractive when congestion decreases. Public transport associations *(Verbundsysteme)* are the talk of the day. *Verbundsysteme* means that cars, buses, railways, subways, bicycles and pedestrians should complement each other. A prerequisite for this would be establishing efficient links between them. As Renner wrote,

> It is time to build a bridge from an auto-centered society into an alternative transportation future characterized by greater diversity of transport modes, in which cars, buses, rail systems, bicycles, and walking all complement each other... A first step governments can take...is to discourage auto use where possible. Local and national governments already impose a variety of physical and financial constraints on automobile use in particular areas or at specific times.[43]

There are technical devices that would allow electronic road pricing. Hong Kong experimented with such a system as early as 1982. There every car is equipped with an electronic number plate that can be read by sensors installed on the roadside. A signal is transmitted from the electronic number plate to a computer center that saves the information. Every car has an account that is indebted when passing certain control points. The owner of the car receives her balance every month. If she does not pay or does not possess a valid electronic number plate, she will be fined. This system allows local authorities to make access to inner areas of cities (or congested areas) more difficult because it becomes costly.[44] Opponents of this system point to possible manipulation of private data and to the fact that financial disincentives favor the rich.

Radical critics of the present system of transportation assume that there are synergistic or holistic relations between the components of this system, i.e. that they reinforce each other. Vehicles with limited range, top speed and acceleration will be ideal for city traffic. Given city administrations' strong interest in cleaner and quieter city traffic, the vehicles might be a strong ally in the electrical network.

Traffic experts often use the term "modal split." This means that different kinds of transport can be done by car, train, ship or plane. According to Vester, a new definition of the modal split should make sure that electrical vehicles are used only for trips shorter than 50 kilometers (which is mainly in the city).[45] The traditional car would be used for trips between 50 and 100 kilometers, the railway for distances between 100 and 1000 kilometers, and the airplane for trips over 1000 kilometers.

In the past, car traffic has grown steadily in volume and intensity. The same is feared for the future. These measures will not be sufficient to accommodate the overwhelming growth in traffic systems and especially car traffic. As Joerges has emphasized, LTS cannot be black boxed for good. They constantly create the conditions for their own transformation. Two typical crises accompany this process, namely a crisis of legitimation and a crisis of control.[46] At present, *downsizing* has become an issue among city planners and traffic experts to tackle both the problem of legitimation and control. I will briefly mention three examples of downsizing.

First, there are attempts to introduce incentives that would encourage new car buyers to purchase a car in the next lower category of horsepower or cubic inches.[47] This measure, which aims to further reduce CO_2 emissions is, however, not considered efficient. Therefore, the German association of automobile manufactures (VDA) sees *upsizing* as an appropriate way to reduce CO_2 emissions by building more roads. The VDA estimates that a better road network will reduce stop-and-go traffic, thereby reducing CO_2 emissions by thirty to fifty percent.[48] Electronic road management is said to reduce them by eight to ten percent and the linking of private traffic and transit systems by five to seven percent. These figures seem to be dubious, with the interests behind this approach too obvious.

Secondly, in some cities, roads are narrowed to slow down and divert traffic in inhabited areas. This measure is linked to reduced speed limits (30 km/h) or to creating a separate bus lane.

Thirdly, many people advocate cities where people live and work in the same areas. Renner put it in the following way:

Reorienting transport priorities can be successful only within the framework of a comprehensive urban policy. There is a symbiotic relationship between land use patterns and transportation networks.... The more concentrated both population and jobs are, the shorter are travel distances, the more mass transit becomes viable, and the more walking and biking occurs. In short, more compact cities foster less individual motorized transport.[49]

It remains questionable if this perspective is compatible with the demands of present day society that reward mobility. Nor does this vision seem very attractive to those who do not want to have their working place next to their home.

Electronic Road Management: The Importance of Standard Setting

Special attention is currently given to electronic traffic guidance systems which are designed to give drivers information about the traffic flow and available parking space. The main problem of traffic guidance systems is finding common standards.[50] Electronic traffic systems are designed to make fuller use of existing traffic capacities, highlighting another feature of technical systems, namely the tendency towards maximum coverage of network activities.

There are several projects under way that are working on establishing electronic road management. As *Automotive Engineering* put it, "Today's cars have very little built-in intelligence for receiving information about the surrounding traffic environment. And most highway systems have little capability for sensing traffic densities and speeds. In other words, we have not-very-smart cars on dumb highways."[51]

In a similar vein, Volkswagen R&D Director Ulrich Seiffert writes on the coupling of different traffic systems. "We still have singular and uncoupled (or only loosely coupled) traffic systems which have all developed independently from each other. The use of electronic media in guiding the traffic between these different systems takes place only in exceptional cases."[52]

Seiffert acknowledges that most trips are undertaken for leisure purposes (see Table 1). The chances for transit systems to break into this domain are not very high since patterns of leisure behavior differ strongly. Therefore, it is important to attract traffic from work, business, and shopping transport. This seems much easier since these activities are characterized by flexible, market-oriented and client-friendly transit systems. Such an approach would relieve cities and regions from congestion.

TABLE 1 Volume of Traffic in Germany.

	Transported Persons (Millions)				Performance (Million Personkilometers)			
	1976	1982	1987	2010	1976	1982	1987	2010
Transit	6 566	6 562	5 788	5 135	68 503	75 951	69 687	66 629
work	1 779	1 411	1 274	840	19 747	15 827	14 490	9 469
education	1 856	2 054	1 655	1 651	14 477	15 130	11 928	11 771
business	121	145	151	155	1 815	2 389	2 463	2 766
shopping	1 472	1 578	1 487	1 370	9 862	13 480	13 256	12 385
leisure	1 332	1 367	1 213	1 106	20 951	26 015	25 345	27 176
holiday	5	8	9	13	1 651	3 110	2 205	3 062
Railway	1 024	1 120	1 088	978	36 250	38 349	39 949	43 267
work	439	461	444	360	7 901	8 021	7 615	6 317
education	217	246	209	191	3 429	3 862	3 170	2 868
business	38	45	58	65	4 016	4 181	5 726	7 997
shopping	128	146	150	143	2 860	3 265	3 121	3 196
leisure	188	208	213	205	14 378	15 319	16 445	18 434
holiday	14	14	14	14	3 667	3 700	3 872	4 455
Car	26 791	27 861	31 180	35 663	419 391	462 402	533 362	638 027
work	6 748	7 695	8 603	9 863	79 270	101 135	116 214	142 735
education	825	1 022	1 002	1 003	9 352	12 698	12 664	13 434
business	4 074	3 702	4 067	4 283	65 138	59 465	67 782	75 005
shopping	5 617	5 672	6 086	7 200	41 465	44 734	49 140	60 142
leisure	9 468	9 704	11 355	13 326	181 817	196 515	238 668	291 369
holiday	59	65	68	77	42 358	47 864	48 894	55 342

Source: Statistisches Bundesamt, DLR, Socialdata München, Berechnungen des DIW, Kloas & Kuhfeld, 1990, p. 360.

In this model, traffic would be relocated from the city to its periphery. Parking in cities would be costly, with scarce parking places, while parking at the periphery would be cheap and offer the possibility of changing to bus and rail.[53]

To avoid park and search traffic, digital radio could be used. The Radio Data System (RDS) has a special Traffic Message Channel (TMC) that delivers digitally coded traffic reports without interrupting current radio programs. With the help of TMC, drivers can be advised of available parking lots, pricing, remaining capacity — and even reserve a parking space. Seiffert also sees using a smart-card[54] that can serve as confirmation for the parking lot, a transit ticket, and a means of payment. Seiffert has strongly lobbied for the car radio as primarily a medium of communication and not so much for the development of new devices. In Seiffert's model, radio and smart cards are technical links that couple different traffic systems.

The president of the *Verband der Automobilindustrie* (Association of Car Manufacturers) Erika Emmerich favors a centralized traffic management that organizes the whole city traffic "under one roof and responsibility" (i.e. individual car traffic, transit, and parking adminstration would be centrally organized). She also has high hopes in the electronic information and steering system as a technical basis for this centralized management.[55]

Safety Aspects

As the previous section has shown, quite a few would-be system builders are on the scene. However, there are some major differences between car traffic and other LTS, such as electricity, telephone, airplane or railway traffic. The differences involve, first, different properties of the system and, secondly, its cultural embeddedness. I will address both of them.

It is believed that the Traffic Message Channel in automobility was modelled after the Traffic Management Coordinator in air traffic. The term "traffic fleet management" suggests such a parallel. However, there seems to be such fundamental differences between air traffic/car traffic and pilots/drivers that one does not have high expectations for success in applying air traffic control principles to car traffic. Dangers inherent in such a solution are not to be underestimated. As LaPorte & Consolini have shown in the example of U.S. air traffic control systems (which they have labelled a "high reliability organization"), there are traffic systems that are complex and tightly coupled but nevertheless exhibit extraordinary safety performance.[56] However, the different components of the overall traffic system do not seem to possess the requirements of "high reliability organizations": individual means of transport are not driven or guided by people who are highly motivated to do a good job within a team. Rather, people like to drive or bike according to their own rhythms, moods and responsibility. Drivers are not pilots and cannot be handled the same way by traffic controllers. Drivers might simply reject the envisioned traffic guidance since they regard it as attacking their autonomy.

What Makes Driving so Thrilling?

This brings me to the second decisive difference that sets automobility apart from other LTS, namely its cultural embeddedness. The quasi-monopoly character of the car traffic system stems from its unique qualities in providing the individual with mobility. As White aptly remarked:

> What if commuters actually like driving to work? A car perhaps represents one of the last bastions of privacy in modern America, where a man is away from his family and his boss and colleagues. He can sing, shout, scratch his ears, turn the radio on loud, and make threatening gestures and shout obscenities at other motorists, all without fear of social rebuke.... A car is responsive to the driver's wishes; it is he who is actually controlling 4,000 pounds of steel and complex machinery. He has control over his immediate environment to a degree probably not equaled anywhere else in his daily routine.[57]

Apart from economic explanations, the car culture has such a stability since it satisfies needs that people project on to it, such as that of belonging, safety, power, comfort, and wholesomeness.[58] Religious and sexual connotations have been especially prominent in critical studies of the automobile. The "abiding love affair" of western cultures with the automobile[59] has manifold reasons which I cannot adequately discuss here. Suffice it to say that ever since the Hollywood dream materialized in Detroit's annual restyled car models,[60] passion has held sway over our mode of private transportation. This passion includes love and hate, and it might not be too far-fetched to compare this relation to religious worship. The analogy not only applies to the worshippers but also to the heretics and dissidents. They all formulate their *credo* in an equally rigorous manner, as is known from religious history.[61]

Many authors have compared the automobile to objects of worship, religious fetishes, and gothic cathedrals. Roland Barthes saw the car as an equivalent to the great gothic cathedrals. According to him, both are big creations of the epoch, projected by unknown artists, and used by the entire population who views them as magical objects.[62]

A simple conclusion awaits these considerations. Since the car traffic system probably mobilizes more passions than any other LTS, the conflicts that emerge around its restructuring will be profound and long-lasting.

Conclusion:
Growth, Stabilization and Decay

It has been said that LTS are designed for expected peak loads, thereby triggering dynamics that can be observed in the history of many LTS, namely the change between phases of high peaks and low valleys[63] (overload and underload). Out of this mutual stimulation a characteristic pattern of growth emerges.[64] Such a pattern consists of slow quantitative growth which later accelerates to a point where growth rates decrease. Now a phase of stagnation or absolute decline sets in.[65]

Intuition would suggest that the car traffic system is a good example for such a pattern. However, it rather behaves differently. The physically embodied network (roads and highways) no longer grows significantly (with exceptions in Eastern Europe); some roads are widened, others are narrowed. The components (automobiles) increase every year, as do the quantitative operations within the system, namely trips e.g. by car, transported persons, and kilometers per person (see Table 2). Earlier estimates that car traffic would decline in the eighties and nineties have been wrong.[66] The introduction of electronic road management will contribute to further growth — or at least delay the trend towards stagnation. The same is true for the potential introduction of electric vehicles.

TABLE 2 Development of Car Traffic, Germany.

	unit\year	1976	1982	1987	2010
Car fleet	Millions	18.9	24.1	27.9	37.2
Total mileage	Billion kilometers	255.0	294.4	356.9	465.1
Average mileage	1000 kilometers	13.5	12.2	12.8	12.5
Average passengers	Passenger per car	1.63	1.55	1.47	1.35
Average trip length	Kilometer	16.0	17.1	17.5	19.2
Transported persons	Millions	25 973	26 665	30 026	33 544
Person-kilometer	Billions	415.1	455.4	526.1	631.1
Trips by car	Millions	10,096	20,536	23,655	28,704

Source: Kraftfahrtbundesamt, Calculations of DIW, Kloas & Kuhfeld, 1990, p. 412.

This analysis confirms Joerges's skepticism about the regularity of a pattern of development within LTS. The predicted upward transformation[67] (which according to him follows from the precarious status of LTS) will lead to a fierce struggle — not only among the producers but also among the general public — in the process of reconfiguration of the car traffic system. The existing entangled alliances in favor of various systems of transport will continue their agonizing struggle. In this quasi-religious conflict, a classical solution from the time of religious wars might be revived, with car traffic banned from certain roads and areas. The slogan *cuius regio eius religio* (whose region, his religion) might change into *cuius regio, eius commeatus* (whose region, his traffic). A separation of different traffic systems can only mean a change in the modal split, with the result of downsizing the road network. (The German comic Karl Valentin once proposed a temporal separation of different elements of road traffic: on Monday only bikers, on Tuesday only cars, on Wednesday only the fire brigade, on Thursday only walkers... In his unique fashion, he extended this scheme from days to weeks, to months, to years, to decades and to centuries.)

Several implications can be drawn from our discussion:

First, in contrast to Hughes' system builder model, the increase of the load factor is an unintentional result of the system's history rather than primarily the design of a system builder or an organization. No single organization runs the car traffic system. The organizations involved often have contradicting interests. There are no system builders in sight who can bring together various dimensions (financial, legal, technical) of the car traffic system and solve critical problems connected to it. Or, in Latour's terms, there are no indications that a new sociotechnical network will stabilize.

Second, the linking of different traffic subsystems is designed to increase the load factor of the whole system. Depending on which version of linking will be finally established, it seems clear that the different subsystems will get tighter coupled than they are at present. In particular, electronic road pricing seems to win more and more political advocates in Germany. The equity argument (see above) is discounted in favor of a "who pollutes will pay" argument, which means taxing actual distances travelled on motorways. Some dangers might be inherent in this change, especially in cases where system components are tightly coupled and complex.[68] Indeed there are traffic systems (such as air traffic control) that are complex and tightly coupled but nevertheless have an extraordinary safety performance. But the car traffic system has so many different characteristics than air traffic that the application of air control technologies would remain ineffective in the best case and in the worst case lead to disastrous results.

Third, it seems safe to say that the automobile industry is pushing along its traditional lines of energy-saving engines, catalytic converters and hybrid engines. Radical innovations based on a different paradigm of car will need a niche in the beginning and a technical and political support network to complement the traditional car system. It is an open question whether a new system will be able to replace the old one. Most probably, there will be an increasing number of traffic systems (including electric cars) which complement rather than replace each other. Changes will occur in quantitative relations among them, a process also influenced by the patchwork of local, regional, national and transnational traffic regulations. How radical the reconfiguration will be depends on factors which cannot be foreseen.

Acknowledgements

I would like to thank the participants of the LTS Conference "Large Technical Systems in Radical Reconfiguration" in August 1992 and in particular Donald MacKenzie for comments on an earlier version of this chapter. Part of the research for this chapter was supported by a grant from the Fritz Thyssen foundation which is gratefully acknowledged.

List of Abbreviations

ABB	Asea Brown Boveri
AC	Alternating current
ADAC	Allgemeiner Deutscher Automobil-Club
AE	*Automotive Engineering*
AEG	Allgemeine Elektrizitätsgesellschaft
ATZ	Automobiltechnische Zeitschrift
BMFT	Bundesministerium für Forschung und Technologie
CO	Carbon monoxide
CO_2	Carbon dioxide
DC	Direct current
DIW	Deutsches Institut für Wirtschaftsforschung
EC	European Community
ETZ	Elektrotechnische Zeitschrift
EUREKA	European Research Koordination Association
EV	Electric vehicle
GM	General Motors
HC	Hydrocarbon

IVHS	Intelligent vehicle highway system
LISB	Leit- und Informationssystem in Berlin
LTS	Large Technical System
MTZ	Motortechnische Zeitschrift
NaS	Sodium sulphur
NO_x	Nitrogen oxide
PROMETHEUS	Program for a European Traffic with Highest Efficiency and Unprecedented Safety
R&D	Research and Development
RDS	Radio-Data-System
RWE	Rheinisch-Westfälische Elektrizitätswerke
SO_2	Sulfur dioxide
TMC	Traffic Message Channel
UCLA	University of California, Los Angeles
VDA	Verband deutscher Automobilhersteller

Notes

1. Flink 1988, p. 315.
2. Hughes, 1983, p. 2.
3. Ibid., p. 5.
4. Ibid., p. 465.
5. Hughes, 1987, pp. 71-2.
6. Ibid., p. 26.
7. Ibid., p. 24.
8. See Schneider, this volume.
9. Ibid., p. 25.
10. Ibid., p. 27.
11. Latour, 1987, pp. 131, 23, 41.
12. Mayntz, 1988, pp. 238-9, my translation.
13. Contrary to what Donald MacKenzie has suggested, I think that load management is not only important in economic terms. True enough, the present car traffic system in Germany does not yield an income. However, the load factor is always present in traffic planning where it has a political dimension. With the attempts at privatization of the highway network in Germany and the introduction of road pricing, the economic dimension might become important as well. See Blüthmann (1993).
14. Seiffert, 1991.
15. Vester, 1990, p. 408 ff.
16. Rommerskirchen, Becker & Eland, 1992.
17. In fact, the electric vehicle technology is as old as the internal combustion engine. In 1899, the electrically powered *Jamais Contente* established

a speed record of 104 km/h. The third main technology was the steam engine (which did not survive as motive power for cars).

18. Vester, 1990, p. 404, my translation. By systems logic Vester means a biocybernetic approach.

19. Vester recommends that the car industry out of its enlightened self-interest takes the initiative for the set-up of a new traffic system. At times, Vester himself looks like a would-be system builder, changing roles from Don Quixote to Machiavelli.

20. Latour et al., 1992.

21. Callon & Latour, 1981; Callon, 1987.

22. The President of Volkswagen, Ferdinand Piëch, admits that such an objective cannot be achieved with engine modifications alone, *ATZ* 94 (1992) pp 20-23. "In addition to engine efficiency, fuel consumption is decisively influenced by the resistance to motion and, most especially, the weight of the vehicle. Weight could be reduced by up to thirty-five percent with the aid of ... aluminium body... With extreme optimization to body aerodynamics, the resistance to motion can be cut by a further thirty-five percent." (id., p. 21).

23. Manufacturers may produce any combination of the first three groups, as long as they meet average fleet emissions each year. Plans for 1998 call for two per cent of each producer's fleet of passenger cars and light trucks to be ZEVs, see California Air Resources Board (1989) and Schäfer (1991).

24. White, 1971, p. 211.

25. Ibid., pp. 216-7.

26. Hård, 1992, p. 75.

27. "Although the relative strength of these trajectories could vary between regions, they are, nevertheless, globally observable." Hård, 1992, p. 67.

28. In the case of tighter emission standards which were introduced in the mid-eighties in the EC, German industry created such an obligatory passage point for the rest of Europe by means of the catalytic converter.

29. It is impossible to list all local projects which have to do with a more ecological car alone, but among them are certainly technical solutions in the field of battery technology, electric propulsion systems, gas turbines, ceramic engines, hybrid technology, range extenders, alternative fuels, solar cell technology, fuel cell technology, cutting the rolling resistance of tires, improving aerodynamic performance, and lighter materials for car bodies and engines. See Bleviss, 1988 for a good overview and also Morales & Storper, 1991.

30. *ATZ* 94 (1992), p. 21.

31. For the term vehicle concept, see Clark and Fujimoto, 1991. Lenz & Wiescholek estimate that this point is reached at a production volume of 100,000 units. Lenz & Wiescholek, 1990, p. 438.

32. Morales & Storper, 1991, p. 35.

33. Lenz and Wiescholek 1990.

34. The rhetoric is so common and ubiquitous that I refrain from giving a specific source.

35. As a manager of ABB Hochenergiebatterie put it, cf. Dustmann, 1989. This gives a hint that proponents of the electric car feel confident to challenge the traditional car on its own battleground.

36. Here the engineers seem to be the mirror image of the sociologists described by Callon and Latour 1981, p. 298. They want to keep out of their world all that disturbs them (the social in the first case, the technical in the second case).

37. *AE*, March 1989: "Volkswagen has completed a major research program which, it says, has demonstrated that lean burn engines cannot match, under *all* conditions, the low pollution levels produced by catalyst-equipped units." (p.74, my italics).

38. *AE*, April 1991, p. 33-36.

39. See Knie, 1991 with respect to the diesel engine.

40. The same is true for the protypes from GM ('Impact') and from Tokyo R&D (NAV = Next Generation Advanced Electric Vehicle), cf. Morales & Storper, 1991, p. 40.

41. For an overview of prototypes of electric vehicles, see Vester, 1990, p. 357-382; Morales & Storper, 1991, p. 37-41.

42. Renner, 1988, p. 25.

43. Ibid., p. 49.

44. See Cerwenka, 1990, pp .76-77 and the cited literature, especially Dawson et al., (1985); see also Blüthmann (1993).

45. This agrees with findings of traffic statistics for Germany where only three percent of all trips are longer than fifty kilometers and twenty-five percent below one kilometer, cf. 'Dokumentation Kontiv 1989' 44 *Internationales Verkehrswesen*, No. 3, (1992), p. 88.

46. See Joerges, 1992.

47. See VDA, 1990, p. 42.

48. Ibid., p. 40.

49. Renner, 1988, p. 51.

50. *AE*, No. 3, (1991) reviews different systems. Standard setting not only touches upon different technical devices but also on health issues. Electromagnetic waves from radiofrequnecies may be harmful for humans. Cf. *Technische Rundschau*, No. 4, 1992, p. 32-34).

51. *AE*, 1991, No.3, p. 15.

52. Seiffert, 1991, p. 553.

53. Parking is a problem as old as the automobile. James Flink gives an illustration: "By the end of 1916 even the editors of *Automobile* were overwhelmed that 'every day in big cities the parking problem grows more acute. ' If it is bad today, and indeed it is so, what will be the situation in three years? We are facing something which was never foreseen in the planning of our towns, a thing which has come upon us so swiftly that there has been no time to grasp the immensity of the problem till we are almost overcome by it.... Articles in *Motor*, for example warned, 'Stop! You are Congesting the Streets,' and asked, 'Will Passenger Cars Be Barred from City Streets?'" (Flink 1975, p. 163).

54. He labels it an anonymous chip card, presumably to forestall criticism of possible violation of data privacy.

55. A step towards this new technology is the European project PROMETHEUS which was initiated in October 1987 as an R&D project within the EUREKA program's framework, see Juhlin in this volume.

56. LaPorte and Consolini, 1990.

57. White 1971, p. 236.

58. See Sachs, 1984, p. 151 ff. It has been pointed out that when Nietzsche thought about the death of God (i.e. in the years 1882 and 1885), Benz and Daimler built their first motor cars. The 'unmoved mover' has died and was resurrected in the mechanical auto-mover. At the peak of melancholic cultural criticism, as marked by Nietzsche, mankind invented one of the most optimistic machines ever seen since it confronts the desire for death with the opportunity of incredible time gain. Conversely, the farewell to the car which is proclaimed in recent years, takes on quasi-religious traits (Schreiber, 1991).

59. Renner, 1988.

60. Armi, 1988; Flink, 1992; Yates, 1984.

61. At least in Germany, it is significant that in many public debates on the issue often not a single well-reasoned statement can be found. This is a hot sphere of ideology.

62. See Barthes, 1957.

63. Hughes, 1983, p. 218 and passim.

64. See Mayntz, 1988, p. 252.

65. Ibid., p. 254.

66. See Kloas & Kuhfeld, 1990, p. 355.

67. Joerges, 1992.

68. Perrow, 1984.

References

Armi, C. Edson. 1988. *The Art of American Car Design*. University Park, Pa.: The Pennsylvania State University Press.

Barrett, Paul. 1975. "Public Policy and Private Choice: Mass Transit and the Automobile in Chicago between the Wars." *Business History Review* 59: 474.

Barthes, Roland. 1957. *Mythologies*. Paris: Seuil.

Bleviss, Deborah Lynn. 1988. *The New Oil Crisis and Fuel Economy Technologies*. New York: Quorum.

Blüthmann, Heinz. 1993. "Die Trasse zur Kasse." *DIE ZEIT* no. 8, (19 Feb.): 13-15.

California Air Resources Board. 1989. *Low Emission Vehicles / Clean Fuels and New Gasoline Specifications — Progress report*.

Callon, Michel. 1987. "Society in the Making: The Study of Technology as a Tool for Sociological Analysis." In *The Social Construction of Technological*

Systems, ed. W. Bijker, T. P. Hughes and T. J. Pinch, 83-103. Cambridge, Mass.: MIT Press.

Callon, Michel & Latour, Bruno. 1981. "Unscrewing the big Leviathan: how actors macro-structure reality and how sociologists help them to do so." In *Advances in Social Theory and Methodology. Toward an integration of micro- and macro-sociologies,* ed. K. Knorr-Cetina and A.V. Cicourel, 277-303. Boston, London and Henley: Routledge & Kegan Paul.

Cerwenka, Peter. 1990. "Überbetriebliche Maßnahmen zur Attraktivierung des ÖPNV in Ballungsräumen - gezeigt am Beispiel von Berlin West." *Verkehr und Technik* 43: 71-81.

Clark, Kim B. and Fujimoto, Takahiro. 1991. *Product Development Performance Strategy, Organization, and Management in the World Auto Industry.* Boston, Mass.: Harvard Business School Press.

Dawson et al. 1985. "Electronic Road Pricing in Hong Kong." *Traffic Engineering and Control* 26: 522-529, 608-615; vol. 27: 13-18, 79-83.

Dustmann, Cord-Heinrich. 1989. "Antriebssysteme für Elektroautos." *Elektrotechnische Zeitschrift:* 48-52.

Flink, James J. 1975. *The Car Culture.* Cambridge, Mass.: The MIT Press.

_____ 1988. *The Automobile Age.* Cambridge, Mass.: The MIT Press.

_____ 1992. "The Olympian Age of the Automobile." *Invention & Technology.* (Winter 1992): 54-63.

Hård, Mikael. 1992. "Technology in Flux: Local Practices and Global Patterns in the Development of the Diesel Engine." *WZB discussion paper* FS II: 92-103.

Hughes, Thomas P. 1983. *Networks of Power. Electrification in Western Society, 1880-1930.* Baltimore: Johns Hopkins University Press.

_____ 1987. "The Evolution of Large Technological Systems." In *The Social Construction of Technological Systems. New Directions in the Sociology and History of Technology,* ed. W. E. Bijker, T. P. Hughes & T. J. Pinch, 51-82. Cambridge, Mass.: The MIT Press.

Joerges, Bernward. 1988. "Large Technical Systems: Concepts and Issues." In *The Development of Large Technical Systems,* ed. R. Mayntz & T. Hughes, 9-36. Frankfurt am Main: Campus.

_____ 1992. "Große technische Systeme. Zum Problem technischer Größenordnung und Maßstäblichkeit." In *Jahrbuch Technik und Gesellschaft,* vol. 6, ed. G. Bechmann & W. Rammert, 41-72. Frankfurt am Main: Campus.

Knie, Andreas. 1991. *Diesel - Karriere einer Technik. Genese und Formierungsprozesse im Motorenbau.* Berlin: Edition Sigma.

Kloas, Jutta & Kuhfeld, Hartmut. 1990. "Status-quo-Projektion des Personenverkehrs in der Bundesrepublik Deutschland bis 2010." Part 1 and 2. *Verkehr und Technik* 43: 355-361, 408-412.

La Porte, Todd R. 1988. "Increasing Reliability in the Midst of Rapid Growth." In *The Development of Large Technical Systems,* ed. R. Mayntz & T. Hughes, 215-244. Frankfurt am Main: Campus.

La Porte, Todd R. & Consolini, Paula M. 1991. "Working in Practice But Not in Theory: Theoretical Challenges of High-Reliability Organizations." *Journal of Public Administration Research and Theory* 1: 19-47.

Latour, Bruno. 1987. *Science in Action. How to Follow Scientists and Engineers through Society*. Milton Keynes: Open University Press.

Latour, Bruno & Mauguin, Philippe & T, Geneviève. 1992. "A Note on Socio-Technical Graphs." *Social Studies of Science* 22: 33-57.

Lenz, Wolfgang & Wiescholek, Ulrich. 1990. "Umwelteffekt und Wirtschaftlichkeit von Elektrofahrzeugen. Ein Bericht für den Deutschen Bundestag über die Förderung von Elektrofahrzeugen." *Verkehr und Technik* 42: 435-443.

Mayntz, Renate. 1988. "Zur Entwicklung technischer Infrastruktursysteme." *Differenzierung und Verselbständigung*, ed. R. Mayntz et. al., 233-259. Frankfurt am Main: Campus.

Morales, Rebecca & Storper, Michael. 1991. "Prospects for Alternative Fuel Vehicle Use and Production in Southern California. Environmental Quality and Economic Development. *Working Paper # 2*." Los Angeles: University of California, Lewis Center for Regional Policy Studies.

Pascal, Blaise. 1921. *Pensées. Oeuvres*, vol. 13. Paris: Hachette.

Perrow, Charles. 1984. *Normal Accidents*. New York: Basic Books.

Renner, Michael. 1988. *Rethinking the Role of the Automobile*. Washington, D.C.: Worldwatch Institute.

Rommerskirchen, Stefan, Becker, Udo & Eland, Mario. 1992. "Entwicklung der Luftschadstoffemissionen des Verkehrs in Deutschland bis 2010." *Internationales Verkehrswesen* 44: 59-66.

Sachs, Wolfgang. 1984. *Die Liebe zum Automobil*. Reinbek: Rowohlt.

Schäfer, Fred. 1991. "Gesetzliche Vorschriften zur Schadstoff-und Verbrauchsbegrenzung bei PKW-Verbrennungsmotoren." *Motortechnische Zeitschrift* 52: 346-355.

Seiffert, Ulrich. 1991. "Verkehr 2000." *Automobiltechnische Zeitschrift* 93: 552-560.

Schreiber, Mathias. 1991. "Abschied vom Auto?" *Frankfurter Allgemeine Zeitung*, (11 May).

Verband der Automobilindustrie. VDA, 1990. *Überlegungen der deutschen Automobilindustrie für ein Gesamtverkehrskonzept*. Frankfurt am Main.

Vester, Frederic. 1990. *Ausfahrt Zukunft. Strategien für den Verkehr von morgen*. München: Heyne.

White, Lawrence. 1971. *The Automobile Industry Since 1945*. Cambridge, Mass.: Harvard University Press.

———1982. *The Regulation of Air Pollutant Emissions from Motor Vehicles*. Washington: American Enterprise for Public Policy Research.

Yates, Brock. 1984. *The Decline & Fall of the American Automobile Industry*. New York: Vintage.

13

Information Technology Hits the Automobile? Rethinking Road Traffic as Social Interaction

Oskar Juhlin

Introduction

When driving along the roads, we constantly have to muddle through situations which demand quick decisions. Should we accelerate, brake or turn? Furthermore, we have to get a clear idea of the immediate environment. Is that junction far ahead? Is that driver going to brake? Considering how many people are out in traffic and the enormous amount of simultaneous actions taking place, it is surprising how well it all functions. Despite its complexity, traffic could be described as a very smooth system with few problems. Drivers and other road users seem highly skilled and very able to participate in this waltz of vehicles.

However, the opposite perspective is just as valid. Instead of focusing on traffic as something self-organized and structured, we could view it as chaotic and uncoordinated. In this perspective, traffic accidents are the focus. Every year, there are thousands of casualties and still more injuries due to traffic. Globally, 300,000 persons are killed in accidents and around 10 million people are injured.[1] Pedestrians are hit and killed by cars. Drivers fall asleep and smash head-on with an oncoming car. Wild attempts at overtaking become fatal.

Consequently, the driver could either be praised for his excellent abilities to participate in the road transport system or scoffed at for his primitive and clumsy command of a dangerous technology. Furthermore, if we focus our attention on the harmonious character of road traffic, we should be looking for the underlying mechanisms that make it work. And if we focus on the breakdowns or system failures, we should be looking for the remedies.

This paper addresses both these perspectives. It sets out to explore the operational system of road traffic, then it analyses the impact of new information technology on that particular system. The organization of the paper is as follows. First, the emergent research field called *Large Technical Systems* (LTS) is explored for conceptual tools that would help analyze a transport system. As an alternative to the macro perspective of LTS, a micro-level perspective of traffic as symbolic interaction is explored with the help of work by W.D. Dannefer and Mary Douglas. Then new conceptions of a road transport system through RTI (Road Traffic Informatics) — as introduced on a large scale by researchers, administrators and car manufacturers — will be explored. Finally, I will interpret the effects of these new technologies on the operation of road traffic, viewed as symbolic interaction.

Looking for a System Operator

In the traffic research field, it is common to define the components of the road transport system as vehicles, drivers and roads.[2] However, such a definition blurs the view of the broader dynamics of this large technical system, i.e. its operation. Cars, roads and traffic as components seem to fit well into the vocabulary of the emergent research field devoted to the study of Large Technical Systems where a distinction is made between vehicle unit, operation and network.[3] In contrast, studies in the traffic research field and others with interests in automobility tend to neglect actual system operation. Usually interest groups and power relations "above" the road transport system — rather than social organization and integration inside the system — are the focus of research.[4]

I will therefore start with a brief analysis of the road transport system as a large technical system. A large technical system is usually described as consisting of a set of elements, both technological and institutional.[5] The institutional element is divided into ownership, organization and rules. Bernward Joerges has specifically pointed at the formal impersonal rules as being an essential part of a technical system.[6]

Most large technical systems (e.g. airways and railways) depend on a central system operator. Can we find a similar feature in road traffic? Traffic planners form a professional group with a centralized position in traffic operation. They could either work through the construction of the network or they could seek to influence traffic operation by trying to favor a so-called modal split, i.e. policies devoted to favor one traffic mode — usually public transport — over the others.

The Australian traffic planner A.D. May has studied the activities of an aggregated traffic system operator, especially those activities that involve traffic-restraining measures. May defines traffic restrains as "those [*measures*] that impose a restriction on vehicle use in order to achieve a significant modification in the mode, time, route or destination of journeys."[7] He identifies four types of control on traffic: (1) physical restrictions, (2) regulatory restrictions, (3) time penalties, e.g. organized congestion with traffic signals, and (4) fiscal control such as road pricing and area licensing.

However, May concludes that "[v]ery few proposed restraining schemes have, in practice, been implemented."[8] Traffic restraint is basically absent in the road transport system. It is apparent that most of these measures, except perhaps organized congestion, are very static and loosely connected to variations in traffic. It is, therefore, highly questionable whether one could talk about a central system operation in road traffic at all.

Peter Kronborg, associated with the Swedish RTI developers, discusses in a recent report the role of a system operator in traffic.[9] He proclaims:

> A major problem is that no one has the responsibility of making road traffic operate in Sweden.... No one is really responsible for the traffic system. Both railways and air traffic have control centers with this responsibility. In Sweden no such road traffic control exists.[10] (author's translation)

As my brief description indicates, I agree with Kronborg that there is not much of a central system operation within traffic. Centralized elements are static and do not correlate to actual events in daily traffic. The traffic sign does not change between peak hour traffic and quieter periods. It does not even care if its messages are followed or not. It is apparent from May's and Kronborg's discussion that we are unable to understand the operation of traffic simply by studying the activities of a traffic control center. However, Kronborg's conclusion that no one is really responsible seems drastic.

A current theme in LTS studies is to focus on operational subsystems, combined with an interest in the rules that govern these systems. Integration between different elements has also been discussed. However, I am critical of directly applying such a perspective to the study of road transport. The LTS studies on railway and airway systems have tended to focus on the macro level. Although large and technical, the road transport system is today very different from other transportation means. The technical integration between the network and the vehicles is very low. There is not even a loosely connected system operator as for example in the air traffic system. The car driver is king of the road with his vehicle as the perfect tool. Thus, an analysis of the driver at a micro level is essential to understand the operational system.

Understanding Traffic Operation as Social Interaction

The empirical focus of this analysis now takes on a new direction as the driver's seat is occupied. Since there are hardly any central system operators in road traffic giving orders (or a technology which automates the process), we must continue our search for the operators elsewhere. It seems that we have to look at the base to find the operators of vehicle units. It is hardly surprising to find that the operators of the road transport system seem to be closely integrated with the act of driving. Therefore, we need to ask ourselves what the drivers do when they drive.

An answer could be to apply the formal rule system, i.e. traffic regulations. But do we understand what happens on the roads by being familiar with traffic laws and how to interpret traffic signs? Empirical investigations seem to reject that explanation. There are formal rules and informal rules used by drivers. Therefore, to strictly follow traffic regulations could be a dangerous driving style.[11] Understanding traffic by investigating its formal set of rules becomes less interesting. These findings also further emphasize road traffic operation as extremely decentralized compared with other transport systems.

The sociologist W.D. Dannefer is an exception to the meager research interest in this field. In the article *Driving and Symbolic Interaction*, a foundation for an understanding of what could be called the operational characteristics of road traffic is outlined.[12] The article focuses on symbolic interaction in traffic, i.e. the process through which drivers communicate with each other to coordinate the movements of their vehicles.[13] Basically, Dannefer distinguishes between situationally based rules and identity-based rules governing the interaction. The former is a universal and impersonal category. The situationally based rules could either be formal (like the traffic laws) or informal. Informal situationally based rules are habitual and acquired through day-to-day participation in traffic, according to Dannefer: "These norms are purely situational, generated by the problems of coordination to which they apply...."[14] For example, Dannefer refers to how cars come to a stop far behind the intersection in order to accommodate a bus making a tight corner turn. Interestingly, the formal rules could be in conflict with informal rules. Norms of courtesy can, for instance, put traffic rules out of order. Another interesting case is when drivers inside a round-about yield to an oncoming car, despite the legal rules giving him the right-of-way. This is explained by Dannefer through the application of a specific informal rule: the flow of traffic makes continued movement an easier, more natural course of action than obedience to both traffic laws and norms of courtesy.[15]

This informal rule will, hereafter, be referred to as the "flow priority" rule. However, Dannefer adds to these impersonal rules a set of rules dependent on personal characteristics. According to Dannefer the latter form of symbolic interaction is specific and based on the categorization of the other actor in socially meaningful terms. The other actor is included or excluded in group solidarity, such as truck drivers' or fast-car drivers' sense of togetherness. When interacting, the driver analyses who the others are. Are they someone to trust? Are they someone to collaborate with? According to Dannefer, there are also certain other features present in an operational categorization (see Table 1).

TABLE 1 Some Operational Categories for Identity-Based Interaction.

Driver-to-Vehicle-Communication	Driver-to-Driver-Communication
• size of the other's vehicle	• age
• aggressive behavior	• sex
• prestige, e.g., price of the car	• dress
	• expression

With these characteristics in mind, the driver analyzes the intentions of the other vehicle and decides which actions to take. However, road traffic is operated more on situationally based interaction than on identity-based interaction. Thus, the impersonal rule system, both informal and formal, is more important than social categories. Dannefer's main argument is that the interactive constraints of the road transport system produce the normative constraints of traffic. In general, traffic is impersonal and anonymous. The feeling of privacy is provided by the contingency of every particular interaction. This also explains, according to Dannefer, blunt rule breaking of commonly accepted rules by drivers performing actions based on, for example, impatient self-interest and sheer expressions of rudeness. There are technological restrictions that affect and sustain the social framework of traffic.

Other sociological studies view road traffic in the same manner. There seems to be a consensus about the dominance of situationally based rules, or impersonal rules guiding actions in traffic.[16] The means of establishing joint action is restricted.[17] Sociologists seem to agree on describing traffic as extremely decentralized, impersonal and based on autonomous drivers.

To some scholars of traffic behavior, this view may even look like a cultural void.[18] However, a definition of culture as consisting only of norms and values is much too narrow and seems to exclude traffic from sociotechnical systems. Social anthropologist Mary Douglas has outlined a typology for analyzing a broad range of cultural manifestations based on the concepts "grid" and "group."[19] This typology of different modes of social control is used to explain how the production of meaning is related to a specific social context. The grid dimension, going from weak to strong, captures the amount of hierarchy in social relations. According to Douglas:

> The term grid suggests the cross-hatch of rules to which individuals are subject in their course of interaction....At the strong end there are visible rules about space and time related to social roles; at the other end, near zero, the formal classifications fade, and finally vanish.[20]

At one extreme of the grid, people are stuck in their hierarchical positions and are restricted to transacting over these varved cultural boundaries. At the weak end of the grid, these restrictions fade away. Instead of being remotely controlled, people live under autonomy and competition. The other concept is that of the group. It is defined "in terms of the claims it [*the group*] makes over its constituent members, the boundary it draws around them, the rights it confers on them to use its name and other protections, and the levies and constraints it applies."[21] In a strong group, a member is constantly reminded of the correct way to behave, what clothes to wear and so forth. Thus, the group dimension captures the amount of solidarity, concord or personal pressures within a group.

In social studies of traffic, theorists agree that traffic is something impersonal and anonymous. It is competitive and restricted by a formal framework which does not have any explicit social connotations: men are allowed to drive like women and the elderly like youth. This description seems to fit well into Mary Douglas' typology. Thus, road-traffic as symbolic interaction is to be understood as a weakgrid and a weakgroup culture.

However, sociologists also seem to agree on the specific technological restraints that characterize the operation of road traffic. The speed and the closed-in position of the driver make communication very difficult. These restraints are now being contested, however, as information technology is being applied to road traffic. In the next section this new technological field will be explored.

Road Transport Informatics — New Tools for Operation?

It seems very likely that road traffic, as we know it, is undergoing a radical reconfiguration, resulting in a much more integrated road transport system. The basic idea is to develop information technology for road transport. The concepts are condensed around a nucleus labelled Road Transport Informatics (RTI). RTI is presented by two principle European research programs, DRIVE and PROMETHEUS. Their expressed goals are to increase road safety and road capacity, decrease environmental pollution and support European industry.[22] Key words in their visions are "integration" and "organization." These technologies seem to induce a radical reconfiguration of the operational system of road traffic.

Ideas on extended information systems for road traffic are not unique to the 1980s. Similar ideas were, for example, evoked at General Motor's exhibition at the New York World's Fair in 1939.[23] However, not until the 1980s was there a boom in developing information technology with regard to road traffic.[24]

In his article "Car Traffic at the Crossroads" in this volume, Reiner Grundmann discusses certain aspects of the field that he labels *electronic road management*.[25] He mentions route guidance and parking management's dependence on coded information within the traditional FM band, as well as the "smart card" development for different charging purposes. However, other telecommunication links are also considered in the RTI programs, such as digital cellular radio and infra-red light. The list of functions under the electronic road management label could also be extended to include demand management, trip planning, automatic policing and integrated traffic signals.

Grundmann's concept of electronic road management, with its new traffic control center, is very appropriate for describing the principle part of the technological development measured in research efforts. Nevertheless, there are other paths in this field, all of them less dependent on globally integrated technologies. There are other means of establishing a connection between the car and its surroundings by using information technology. Basically, there is a distinction between telecommunications technology and computer vision technology. There is also a distinction between different means of telecommunication and the elements they may connect. Here a great difference can be seen between different forms of short-range communication and, as already mentioned, global links such as coded-radio information and digital-cellular radio. Reiner Grundman briefly touches upon these other directions in the RTI field by presenting their project names, i.e. *Improved Driver Information*, *Active Driver Support* and *Cooperative Driving*. These projects are not based on global communication links providing a channel between a centralized-management center and a car. Instead, they use either computer-vision technology or so-called short-range telecommunication.

The projects *Active Driver Support* and *Improved Driver Information* are related to developing computer sensing and computer monitoring. Most of the attention has gone toward technologies that facilitate the driver's task. For example, obstacles in the trajectory of the car's path could be detected by use of such phenomena as artificial intelligence, artificial vision, infra-red light or microwaves.[26] Microwave radars could be supported by transponders which reflect, amplify and modulate the signal.[27] Transponder-supported microwave radars could be used to communicate from a vehicle to the roadside or between vehicles. This is one example of so-called short-range communication that is drawing a lot of interest today. With short-range communication, a link is established between two elements in proximity, e.g. between two cars, a car and a traffic signal, a car and a traffic sign or a car and a roadside beacon or antenna. Obstacle detection could also be achieved with microwaves and transponders placed on children, the elderly or road workers.[28]

The information on vehicles' positions, drivers' status, car status, traffic signs and road topology could either be presented to the driver purely as information or connected to a vehicle control system and, thus, automating some of the driver's tasks. The new technology would then start to play an active role in automobile driving. The development of cruise-control systems (already in use) is seen as a first step in the direction of automatic driving.[29] Accidents caused by cars equipped with cruise control crashing into the back fender of a slower car could be avoided, according to the engineers.

The concept of *cooperative driving* is often regarded as the most visionary and revolutionary in RTI development. It refers to functions in which more than one car is equipped with new devices and some kind of computer-supported coordination is conducted vis-à-vis either other cars or vis-à-vis traffic signals. With regard to telecommunication between vehicles, information generated by one car could be communicated to other fellow drivers. The information provided from a friction-detection sensor could, for example, be sent to oncoming traffic. This automated politeness is perhaps the most beautiful idea in these programs: Take care! It is slippery.

Lane-changing and overtaking will be increasingly automated, according to the PROMETHEUS program. A technology based on direct telecommunication between drivers sends status information from one car to the others, making it possible to either warn the driver or automatically adjust the vehicle when critical situations occur.[30] Moreover, the automatic control systems could be used to form condensed high-speed convoys of cars. This is perhaps the most remarkable idea in connection with RTI technology.[31]

To summarize this exposé, we can conclude that there is a vast field of engineering devices emanating from the notion of Road Transport Informatics. If we distance ourselves from the project texts, we could question the description of this technology. What is RTI really about? A clue is given in a video that was launched by the PROMETHEUS program in 1986 to promote its ideas:

> If the automobile is really going to fulfill its role as the vehicle of individual travel and symbol of freedom, it is going to mean unprecedented safety and *organizing flow* [my italics] for maximum efficiency and convenience.[32]

As indicated earlier, these programs could be interpreted as a way to rescue and extend the car industry. Information technology could perhaps reconfigure the existing road transport system to overcome the constraints it experiences today.[33] It seems clear that organization flow is a key concept in the development of these new technological artifacts. Road Transport Informatics promises to reconstruct the road transport system, possibly radically. Thus, the next step in analyzing the impact of these new devices is to apply the conceptual base of present traffic — widely ignored both in engineering texts and in non-engineering academic studies — to traffic as social organization.

Radical Reconfiguration of Driver Integration?

I suggest that the fields represented in RTI development — computer sensing and computer monitoring (also refered to as *Improved Driver Information* and *Active Driver Support*), cooperative driving and road management — embody three different trajectories for the organization of the road transport system. Computer sensing and computer driving form a conventional trajectory in which the driver on his own will be responsible for making the decisions. The car is seen as an autonomous unit.[34] Thus, the major characteristic described by Dannefer will still be that traffic is impersonal and anonymous.

However, two other fields — electronic road management and cooperative driving — are both rejections of the business-as-usual model. Electronic road management, provided through the communication links of modern technology, creates a new position in the operational system of traffic. This is a development where the organization of the operational subsystem becomes more centralized. Cooperative driving, on the other hand, links autonomous drivers to each other and constructs a new type of relationship and a sense of dependency at the local level. Thus, three paths or trajectories are visible for drivers' integration in road traffic as a cultural system, depending on different foundations for its social order: (1) autonomous driving, (2) centrally dependent driving and (3) locally dependent driving.

Some Elements in Centrally Dependent Driving

A number of functions in the RTI field require a control station of some kind outside or above traffic. This control station is called a traffic management center. It is basically a new feature in road traffic.[35] The actors behind the construction of this new and central position in the road transport system must decide on the rules or strategies under which the management center should operate. Here we have two main directions that are recognized among RTI actors. Ian Catling, an influential manager of RTI development at the European level, presents both: "There will be a difference between optimizing for individual route selection...and for an overall community based optimum."[36] Here the question of an active system operator is brought up.

Furthermore, the new system operator could choose to apply different strategies within the two main directions. The most passive form would be to *broadcast* most of the information generated and permit the driver to do the analyzing. The next strategy would be to give *recommendations* to the driver. And finally, the system operator could give *orders*.[37]

It is essential not to jump to the conclusion that electronic road management implies giving orders based on global system optimization strategies. Such a system would probably meet too much resistance from car drivers. However, road management in its weakest form (i.e.

providing global traffic information) implies a shift in organizational structure. New dependencies will arise between drivers and this new outside position. Furthermore, the strategies of a central-system operator may not only aim at optimizing road traffic according to some higher level of drivers' interests or their "volonté générale." The strategies could also be guided by the interests of, for example, pedestrians and residents.

Cooperative Driving — Locally Dependent Driving?

Another way to reconfigure the road transport system is to increase the integration between the vehicle units, constructing dependencies between drivers at the local level. RTI could be able to provide a new mode for communication between cars and drivers. It is, thus, not primarily a globally integrated system, but increasingly integrated at the local level.

Apart from the communicative character of driving, the technology of *cooperative driving* is seen as a way to automate information gathering from the surroundings of the car — or just a form of increasing information available to the driver.[38] The two-way telecommunication between vehicles is presented as nothing more than a new tool for the autonomous driver. However, the notion of cooperative driving is more than that.

As noted above, Dannefer and other sociologists have pointed out the situationally based interaction and the dominance of impersonal rules governing traffic. With the new technology of *cooperative driving*, new instruments and more time will be available to the driver for the interaction in question. Symbolic interaction could then be extended and this could be a base for a new form of car language.

The application of Dannefer's classification of different forms of operation opens up a discussion concerning the impact of cooperative driving on the configuration of the road transport system. Decisions derived from impatient self-interest or expressions of rudeness differ from acts made out of courtesy rules or flow-priority rules, although all these actions could be in conflict with traffic regulation. Thus, the "naughty" drivers could be classified as non-collaborative and aggressive when breaking common rules and norms while the "good" ones could be viewed as collaborating. The latter follow the norms of joint action. Cooperative driving could change the balance between two positions.

A somewhat more opportunistic interpretation is to exaggerate a supposed consensus — or the will to cooperate — among drivers and disregard the problem of aggressive driving. The drivers want to collaborate but are hindered by the technology of today's automobiles. Increased information about the others' intentions will increase cooperation and joint action. Thus, there is consensus and harmony in traffic which is hindered only by lack of information. If inter-vehicle telecommunication is used, there will be spontaneous interaction and joint action. New technology such as RTI would be an answer to this problem and it would not affect the drivers themselves in any way. The drivers' behavior or culture are not included in this calculation. This way to portray the driver — rubbing them the right way — is, of course, in line with the interest of the car manufacturers and their desire to keep on the side of their customers.

However, there is another way to interpret the impact of cooperative driving based on the analysis of traffic culture as an autonomous, competitive, weak grid and weak group culture. Cooperative driving is then much more than a technological solution to a universal driving problem. It is also an idea of changing traffic culture and is seen as an instrument to construct not only something material but also something cultural. Drivers are not left alone. Their conduct, defined as non-collaborative and problematic, has to be changed to overcome the problems or reverse salients (Hughes 1983) of the road transport system. The new technologies used to communicate between cars will then create cooperation between today's autonomous drivers.

This second, non-opportunistic view of RTI as a way to discipline drivers comes from the assumption that traffic culture is much different from the rest of society: it is extremely aggressive and competitive. If inter-vehicle communication or communication in a general sense is established, traffic culture will perhaps conform to the rest of society and not only fulfil the needs of a driving monad. Drivers will no longer allow themselves to punish each other brutally or break common rules as they do today. What specifically makes society's exterior traffic more stable is then less interesting than the material restrictions on it. The disciplining mechanisms could then be said to be be grounded in society's rules and regulations.

On the basis of Dannefer's description of the operational system and Mary Douglas' typology of social organization, I outline two possible mechanisms for such a radical reconfiguration of the autonomous and competitive social framework of driving.[39] The operational principles will in both cases be changed due to increased identity-based interaction. Drivers will not only be supplied with more time for interaction, they will possibly also get new means for symbolic interaction. Ultimately RTI could provide a two-way channel for conversation between cars. By using the new car language spoken through the new devices, the drivers could increasingly categorize and treat each other socially.

Returning to Mary Douglas, we see two ways for such reconfiguration of the operational system. First, a social stabilization could be achieved either by moving towards stronger grid relations, increasing the role of social hierarchies in traffic, or by moving towards group dynamics that provide tougher and more solid norms. Strong grid stabilization based on personal interaction will allow a stabilization based on social structures. Social hierarchies that guide our lives outside of the traffic culture could then guide and discipline driver behavior. In traffic interaction, e.g. over-taking or turn-taking, drivers negotiate about which individual driver should yield. This could increasingly be affected by gender, class, age, wealth, etc.

The second type of reconfiguration of the operational system will be that increased identity-based interaction could lead to stronger normative constraints and formations of social groups in traffic. To a greater extent, drivers could determine if the approaching vehicle was controlled by a "real" driver, i.e. within his own cultural group. Or early in the process of interaction he could recognize him as "alien." As Knapper and Cropley (1981) conclude, identity-based interaction is intertwined with impersonal rule following. The adherence to specific sets of rules is also dependent on identity-based interaction.[40]

In a conflicting situation between cars, the drivers must, for example, decide whether to follow the impersonal rules or not. With RTI, the driver gets more devices with which to determine if the other driver is a rulefollower who is searching for "joint action." Information on the interaction could afterwards be sent to other cars in the vicinity. The group of rule followers will then have more information on "naughty" drivers and could more easily exclude them from group collaboration. The impatient, self-interested driver meets more resistance from drivers who know what he is up to. Majority rules or norms would then perhaps have a stronger influence on car driving.

It can, thus, be concluded that RTI may be able to discipline aggressive and self-interested drivers in a common traffic culture. This could, of course, have a positive effect on traffic safety. However, on the whole it is hard to predict RTI's impact. For example, a problem that causes accidents is the difficulty drivers have establishing "joint action" where there is a choice between a multiple set of rules.[41] RTI devices might even make this list longer, especially when local cultures in specific regions and towns could influence new car languages.

Conclusion

In the last six years there have been three international conferences devoted to the dynamics of so-called Large Technical Systems. The interest in this field of research has been focused on telecommunication, energy systems and transport systems such as railways and aviation. However, there has been no attention devoted to the road transport system, even though Bernward Joerges' definition of a large technical system does not seem to exclude such a study. Large technical systems are, according to Joerges:

> those complex and heterogeneous systems of physical structures and complex machineries which (1) are materially integrated, or "coupled" over large spans of space and time, quite irrespective of their particular cultural, political, and corporate make-up, and (2) support or sustain the functioning of very large numbers of other technical systems, whose organization they thereby link. [42]

Maybe the negligence to study road transport systems could be explained by an implicit difference between large technical systems analyzed and the road transport system. Joerges gives us a clue to the difference when exemplifying some LTSs. He mentions *"integrated* [my italics] transport systems, telecommunication systems, water supply systems, some energy systems...."[43] This seems to imply that a large technical system has to be integrated, or tightly coupled, to qualify as an LTS. What are then the characteristics of an integrated transport system? In the first anthology on Large Technical Systems, a number of studies are presented.[44] These studies reflect the following characteristics of LTS:

- Integration between vehicle and network by information technology, e.g. the telegraph.
- Integration of organizations, e.g. the railroad organization, with no company split between building, maintenance and operative functions.
- Centralized organizations by the invention of the modern corporate management structure.

As we have seen, the road transport system is extremely decentralized. If one intended to describe the operational system, a switch of perspectives must take place, namely away from large organizations towards the activities of individuals. This is quite apparent when studying traffic.

Furthermore, the methodological implication (i.e. trying to focus on the individual and the social framework in relation to a large technical system) could be a way to approach even highly centralized and integrated LTSs. If a reconfiguration of the traffic system occurs, the driver will probably become more socially integrated and accordingly follow the common rules more strictly. But this is a process — and something that cannot be taken for granted. Thus, the question of how the social integration of individual operators in other LTS was established should be raised.

With the introduction of RTI there is a possible shift that will perhaps increase the importance of organized actor groups in operating road traffic.[45] To sum it up, it is possible to say that one likely outcome of the construction process is a more integrated road transport system. This reconfiguration will perhaps have radical effects on the components of the system, specifically the technical and institutional core.

Notes

1. Weaver, 1987, p. 48-50.
2. Lay, 1981.
3. Joerges, 1988, p. 25 and Caron, 1988, p. 72 and La Porte, 1988, p. 223 and Salsbury, 1988, p. 38.
4. Irwin, 1985 and Tengström, 1990.
5. Kaijser, (forthcoming).
6. Joerges, 1988, p.24-25.
7. May, 1986, p. 109.
8. Ibid.
9. Kronborg, 1991
10. Ibid., p. 4.
11. Knapper and Cropley, 1981.
12. Dannefer,1977.
13. See also Swan and Owens, (1988), Goffman, (1971), Boltanski, (1975) Rumar, (1988), Marsh and Collett (1986), Bliersbach and Dellen, (1980), Turnbaugh and Turnbaugh (1987) and Richman, (1972).
14. Ibid., p. 35.
15. Ibid.
16. Dannefer, 1977, Goffman, 1971 and Richman, 1972.
17. *Ibid.* and Swan and Owens, 1988, Rumar, 1988, Knapper and Cropley, 1981.
18. Rumar, 1988.
19. Douglas, 1982.
20. Ibid., p. 192.
21. Ibid., p. 191.

22. Commission of the European Communities, 1988 and PROMETHEUS, *PROgraMme for a European Traffic with Highest Efficiency and Unprecedented Safety* (Stuttgart).

23. Juhlin, 1991b.

24. Juhlin, 1991a.

25. Grundman, this volume.

26. Blosseville, Sellam, et al., (Brussels, 1991) p. 714-15, Alvisi, Deloof, et al., (Brussels, 1991) p. 945-46 and Palmquist, 1991.

27. Palmquist,1991.

28. *CAROSI* , 1988, Rumar, 1987, p. 39.

29. Karlsson, Schüssler et. al., 1990, p. 25 and Palmquist, 1991.

30. Video, PROMETHEUS programme 1989, and Beadle, Chanine et al., 1991.

31. Ibid and ARISE, 1985.

32. Transcript PROMETHEUS-video (1986).

33. Agnelli et al., 1990.

34. Karlsson, 1990b.

35. PROMETHEUS, 1989.

36. Catling and Bell, 1990.

37. Mangematin and Callon, 1991.

38. Ibid.

39. I have found such a discussion in conversations with actors, although not formulated explicitly.

40. Knapper and Cropley, 1981.

41. Ibid.

42. Joerges, 1988, p. 24.

43. Ibid.

44. Mayntz and Hughes, 1988. I am referring to the contributions by Salsbury, Caron, Heinze, Heinrich Kill, Galambos, and La Porte.

45. Juhlin, 1992.

References

Agnelli, U., Johansson, G. L. et al. *The Challenge of Road Transport Telematics*. Brussels: DRIVE Strategic Consultative Committee, 1990.

Alvisi, M., Deloof, P. and Linss, W. "Anticollision Radar: State of the Art." In *Advanced Telematics in Transport*. Brussels: Elsevier, 1991.

Beadle, P., Chanine, M., Kremer, W. and Votsis, G. "Vehicle Inter-Communication: Architecture and Perspectives for Implementation." In *Advanced Telematics in Transport*. Brussels: Elsevier Science Publisher, 1991.

Bliersbach, G. and Dellen, R. G. "Interaction conflicts and interaction patterns in traffic situations." *International Review of Applied Psychology* 29(1980): 475-89.

Blosseville, J. M., Sellam, S. and Guillen, S. "Automatic Incident Detection Using Computer Vision Techniques." In *Advanced Telematics in Transport*. Brussels: Elsevier, 1991.

Boltanski, L. "Les usages sociaux de l'automobile: concurrence pour l'espace et accidents." 1975 March: 2.

Caron, F. "The Evolution of the Technical Systems of Railways in France from 1832 to 1937." In *The Development of Large Technical Systems*, eds. R. Mayntz and T. P. Hughes. Frankfurt am Main: Campus Verlag, 1988.

CAROSI. Car Roadside Signalling. SAAB and Volvo: 1988.

Catling, I. and Bell, M. G. H. *The Role of RTI in Combatting Congestion.* 1990.

Commission of the European Communities. *DRIVE Workplan.* Brussels: 1988.

Dannefer, D. W. "Driving and Symbolic Interaction." In *Sociological Inquiry* 47, no 1 (1977): 33-38.

Douglas, M. *In the Active Voice.* London: Routledge & Keegan Paul, 1982.

Goffman, E. *Relations in Public: Microstudies of the Public Order.* Penguin Press, 1971.

Grundman, Reiner. "*Car Traffic at the Crossroads: New Technologies for Cars, Traffic Systems, and their Interlocking.*" This volume.

Irwin, A. *Risk and the Control of technology: Public Policies for Road Safety in Britain and the United States.* Manchester: Manchester University Press, 1985.

Joerges, B. "Large Technical Systems: Concepts and Issues." In *The Development of Large Technical Systems, ed.* R. Mayntz and T. P. Hughes. Frankfurt am Main: Campus Verlag, 1988.

Juhlin, O. *Prometheus bakom ratten: En social omkonstruktion av vägtransportsystemet.* Borlänge: Swedish National Road Administration report 1991:26, (1991a).

_____: "The Car as a Tool or a Component." In *Tema T at CRICT*, London: Centre for Research into Innovation, Culture & Technology Discussion papers, 1991b.

Kaijser, A. *Staten och infrastrukturen: Historiska erfarenheter och framtida utmaningar.* Stockholm: Carlssons förlag, forthcoming.

Karlsson, T., Schüssler, R. et. al. *Early IRTE Scenario.* Brussels: DRIVE SECFO Deliverable, 1990.

_____. *PROMETHEUS-Ett EUREKA-program.* Mariehamn, Åland: 1990b.

Knapper, C. K. and Cropley, A. J. "Social and Interpersonal Factors in Driving: Progress in Applied Social Psychology." In *Progress in Applied Social Psychology*, eds G. M. Stephenson and J. M. Davis. New York: John Wiley & Sons, 1981.

Kronborg, P. *Störningar i vägtrafiken.* Stockholm: Transport Research Institute report 1991:13.

La Porte, T. "The United States Air Traffic System: Increasing Reliability in the Midst of Rapid Growth." In *The Development of Large Technical Systems*, ed. R. Mayntz and T. P. Hughes. Frankfurt am Main: Campus Verlag, 1988.

Lay, M. G. "*Roads into the future: Designs for the evaluation of road traffic.*" Proceedings of Road Planning and Prioritizing. Stockholm: International Road Federation, 1981.

Mangematin, Vincent and Michel Callon. "Technological Competition, Strategies of the Firms and the Choice of the First Users: The Case of Road Guidance Technologies." Paper from Colloquium on Management of Technology: Implications for Enterprise Management and Public Policy, Paris, 1991.

Marsh, Peter and Peter Collett. *Driving Passion: The Psychology of the Car.* London: Jonathan Cape, 1986.

May, A. D. "Traffic Restraint: A Review of the Alternatives." In *Transpn.Res.*, 20A, no. 2, (1986): 109-121.

Mayntz, R. and Hughes, T. P., eds. *The Development of Large Technical Systems.* Frankfurt am Main: Campus Verlag, 1988.

Palmquist, U. *Volvos Autonomous Intelligence Cruise Control*: Short Technical Report, PROMETHEUS S/IT4:EPTS, 1991.

PROMETHEUS. *Programme for a European Traffic with Highest Efficiency and Unprecedented Safety.* Stuttgart.

PROMETHEUS. *"FUNCTIONS" or how to achieve PROMETHEUS "Objectives."* Stuttgart, 1989.

Richman, J. "The Motor Car and the Territorial Aggression Thesis: Some Aspects of the Sociology of the Streets." *Sociological Review* 20 no. 1 (1972): 5-27.

Rothengatter, J. A. and Harper, J. "The Scope and Design of Automatic Policing Information Systems with Limited Artificial Intelligence." In *Advanced Telematics in Transport.* Brussels: Elsevier, 1991.

Rumar, K. *Social Impact/Acceptance: Definition phase.* Stuttgart: PROMETHEUS, 1987.

_____. *The Role of Human Behaviour: Psychological Aspects.* Hamburg: European Conference of Ministers of Transport, VTI Medd. 89.0079, 1988.

Salsbury, S. "The Emergence of an Early Large-Scale Technical System: The American Railroad Network." In *The Development of Large Technical Systems,* eds. R. Mayntz and T. P. Hughes. Frankfurt am Main: Campus Verlag, 1988.

Swan, A. L. and Owens, B. M. "The Social-Psychology of Driving Behaviour: Communicative Aspects of Joint-Action." *Mid-American Review of Sociology* 13, no. 1, (1988): 59-67.

Tengström, E. Bilsamhället: Bilens makt och makten över bilismen: *Miljö Media Makt* , ed. S. Beckman. Stockholm: Carlssons, 1990.

Turnbaugh, William A. and Sarah Peabody Turnbaugh. "American Greetings: Hand Signalling on the Highways." In *Symbolic Interaction* 10 (1987):1.

Weaver, P. M., ed. *Living with the Automobile.* Borlänge: Swedish National Road Administration, 1987.

Reflections on Systemicy in Technology

14

On Systemic Technology

Svante Beckman

Background and Purpose

Recent research into large technical systems (LTS) is oriented towards several concerns. Three major ones are prominent. One is empirically focused on extensive systems of physical infrastructure such as electric distribution systems, railroad systems, and telecommunications systems which reflect a larger concern with industrial dynamics, as well as with exploring the logic of this class of organizations. Another is conceptually focused on developing a systems approach to the social and historical study of technology as a strategy for integrating the study of technology into the social sciences. A third concern is with systemicy and the alleged modern growth of systemicy in technology as highly problematic in view of human control, assuming that increasing systemicy connotes things like growing momentum, rising complexity, increasing global interdependence and a growth of autodynamics in technical change.

Related to these overlapping concerns are different senses of the shared notion of "system." In relation to systems of technical infrastructure, "system" is a big complex sociotechnical thing "out there," a kind of readily discernible organization that may be treated as an empirically given object of study. In relation to a systems approach, "system" is rather a way to conceptualize technology, as well as other things, as interactive wholes. It is a way of ordering observations about technology and a way to construct empirical objects of study. The system is thus "in here" rather than "out there." In relation to a concern with growing systemicy in technology, "system" is neither a class of organizations nor a principle for ordering observations about technology. Instead it connotes a variable structural property of technology, including sociotechnical organizations, of being more or less systemic. It also connotes a transformation of regulation in technical change towards a principle of systems regulation to which one may refer when explaining technical change.

In the common background of these conceptualizations of systemicy in technology, there are three general questions. The first one concerns what is systemic about anything, including technology. Here the dimensions and eventually the measures of systemicy are in focus. The second question is about what it is in technology that "interdepends" in a systemic way. In focus is the basis of systemicy in technology. The third question relates to the general meaning of systems regulation of technical change. Here the basic problem is the definition of systems regulation in relation to alternative principles of regulation of technical change.

This essay deals with these three questions. I will not so much answer them as offer three conceptual tools for developing the answers. In relation to the meaning of systemicy in general, I will start from Herbert Spencer's classic definition of evolution in terms of differentiation and integration and turn it into a conceptual map of dimensions of systemicy. Concerning what is basically systemic in technology, I will start from Rosenberg's notion of "technical imbalance" and Hughes' richer concept of "reverse salient." Both can be interpreted as conceiving of "operations" as that which basically is systemically interdependent in technology. From this line of thought, I will propose a classification of general types of operations in three dimensions which are useful in identifying the general nature of systemic interdependence in technology. In relation finally to the question about the nature of systems regulation and its alternatives as a regulative principle for technical change, I will propose an ideal-typical, two-dimensional mapping of principles of social regulation. This map identifies four basic types of social regulation of which systems regulation is one. It provides guidelines for appreciating changes in regulation of technical systems as well as insights into the basic theoretical conflicts over how to explain technical change.

Though this essay concerns matters in the conceptual backyard of LTS studies, it aspires to be of some use for empirically oriented students. At least on a general level given the variety of ways in which "system" can be applied to the field, a large technical system may sometimes neither be very systemic nor may its change require explanations in terms of principles of systems regulation.

Systemicy — A Spencerian Approach

While there are plenty of loosely worded definitions of "system," we lack a clear conceptual framing of systemicy as a variable structural property of things studied, including technology.[1] Most studies of LTS show or imply that these organizations have grown historically from smaller and less systemic origins. Some students believe that technology is growing more systemic, connoting increasing interconnectedness, complexity, size, ubiquity, determinacy, autodynamics, and momentum.[2] To clarify such claims, one obviously needs a conceptual device that at least in principle allows us to measure degrees of systemicy. Herbert Spencer's conception of what evolves in evolution is here a point of departure since it — with some revisions — lends itself to a broad definition of systemicy. In this I will ignore his grand theory of a universal trend towards increasing levels of differentiation and integration in biological and cultural units (progressive evolution) as well as his identification of this trend with general progress.

Unlike most modern evolutionists, Spencer believed that evolution is progressive. His definition of evolution is designed to capture what progressively evolves in evolution and to pinpoint the basic dimensions of this kind of change. Evolution to him is a process of increasing differentiation coupled with a process of increasing integration. This interaction, driven by the struggle for existence, gives rise to forms of life and culture representing higher degrees of what he eventually called "determination," meaning higher levels of complexity and specificity of identity, higher degrees of self-determination, and a greater versatility and efficiency in strategies of survival in biological and cultural units. Increasing "determination" means that the relation between a unit and its environment changes in such a way that its modes of adaption become increasingly determined by its internal processes and increasingly selective in its response to environmental change.[3] The mechanism responsible for rising determination is roughly one of functional interdependence, meaning that higher levels of differentiation in a unit permit and require higher levels of integration and *vice versa*.

The relevance of this line of reasoning for the concept of systemicy is quite obvious on a general level. It is undeniable that the idea of a system essentially involves the notion of an integrated whole of some differentiated parts. Spencer's conception of changing levels of determination could equally well, or even better, be understood in terms of levels of systemicy and any processes of systemization must indeed involve the interplay of an integrative and a differentiating process.

To make use of Spencer, however, we must first improve upon his idea. His key concepts of integration and differentiation are simply too broad. Also, they are not well separated, with the consequence that some central features of systemicy are ambiguous cases of both differentiation and integration. This applies most clearly to "hierarchy," meaning, for instance, the number of control levels in a machine or an organization. Hierarchy signifies both a form of integration and differentiation. Integration may also ambiguously refer either to system inclusiveness, i.e. the number of elements integrated in it, or to system coherence, i.e. the strength of connections between elements. Accommodating these distinctions while trying to keep conceptual complexity down may generate the following identification of dimensions:

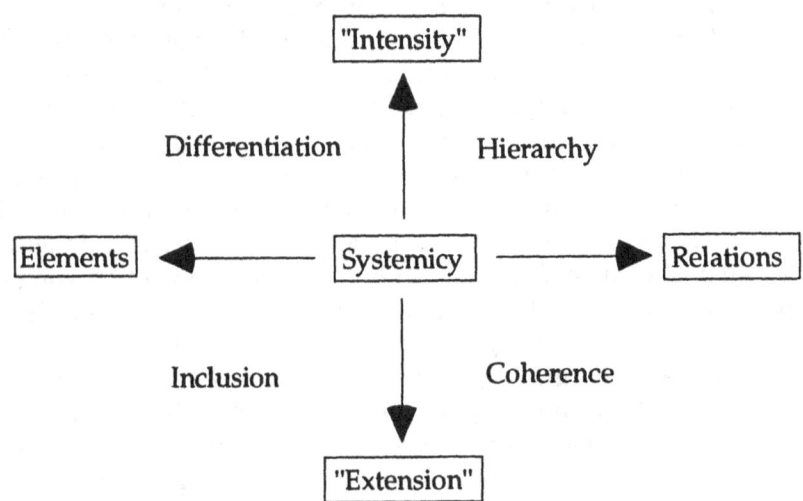

FIGURE 1 Basic dimensions of systemicy.

The figure names four interdependent master dimensions of systemicy which are at least in principle measurable. They are identified with the help of a simple two-dimensional conceptual map and are distinguished between what can be roughly referred to as "intensity" and "extension" on the one hand, and between a reference made to system elements and to system relations, on the other. The overall systemicy of a unit is a function of its values in these four dimensions: how many elements are included in the system (inclusion), how many different kinds of elements it contains (differentiation), how cohesive the relations between elements are (coherence) and, finally, how many levels of hierarchy there are.

The precise meaning of the four dimensions of systemicy differs according to the nature of the system they are used to describe. For example, when applied to a body of scientific knowledge, hierarchy stands for the levels of abstraction it contains, differentiation for the number of concepts it employs, inclusiveness for the volume of facts it accounts for, and coherence for the degree of compatibility between propositions belonging to it. When applied to a business organization, hierarchy connotes levels of decision, differentiation stands for the variety of functions that its activities serve, inclusiveness connotes its size in terms of the volume of contained activities, and coherence for the strength of the ties between activities. The meaning of these concepts when applied to technology will be shown below.

The general nature of interdependence between the four dimensions of systemicy is supposedly the one Spencer assumed: change in any one of them requires and permits changes in the others. Implied here is a principle of appropriate composition of systemicy, which may be defined in terms of the successful persistence of the system, i.e. if the four aspects are not minimally adapted to each other, then the system will not persist. It is thus in this functional, rather than causal, sense that changes in any one of the four aspects may be said to require and permit adaptive changes in the others. The principle implies that some changes observed in systems may be interpreted as internal adaptions of one aspect of systemicy to previous changes in other aspects.

Here we can pinpoint a notorious ambiguity in Spencer's proverbial notion of evolutionary fitness ("the survival of the fittest"), covering as it does both internal aspects — the appropriate composition of a system — and the external aspect — its environmental adaption. It follows that changes observed in a system may either be read as adaption to previous internal changes of systems' properties in view of the principle of appropriate composition, or as adaptions to previous change in system environments (or as a mix of both). In connection with this, we clearly need not only a principle of appropriate composition of systemicy but also a principle of appropriate level of systemicy in terms of environmental "fitness." Changes in the systemicy of a unit may thus be explained in terms of environmental requirements and/or permission to increase or decrease systemicy. As an example, it seems that the level of systemicy attainable in business organizations is kept down by high rates of innovation in their environments. Too high levels of systemicy impinge on their adaptive success. Similarly it is sometimes argued that for some types of LTS, rising systemicy exposes them to increased risks of large scale failure.

At a schematic level of analysis we may assume that the four sub-variables making up the compound variable "systemicy" move together. Rising levels of differentiation and inclusiveness in for example a piece of machinery or a business company go hand-in-hand with rising levels of hierarchy and cohesion. This may guide attempts to explain changes in systemicy in the sense that the addition of levels of control in such units is explained as a functional adaption to increased degrees of differentiation. When elaborating on this line of analysis, the picture becomes more complicated. Given the logical construction of my concept of systemicy, the same level can obviously be attained by combining values of the subvariables in different ways. Assuming, for example, two systems to be equally systemic and to be equally differentiated and hierarchical, one system may attain this level by combining a relatively higher level of coherence with a lower level of inclusiveness as compared to the other system. A big, loose system may thus equal a small, tight one. This opens up the possibility for developing typologies of systems according to their profile of composition, as well as for an analysis focusing on changes in the composition of their systemicy rather than on changes in their level of systemicy. The present wave of deregulation in established infrastructural LTS seems to involve both types of change.

One might consider "complexity" as another term to label overall systemicy. If used as a common language notion for units with increasingly differentiated elements and increasingly differentiated internal relations, then rising systemicy goes hand-in-hand with rising complexity. If complexity is used in the sense of information theory — the number of "bits" required to describe a phenomenon — then rising systemicy typically keeps complexity down. For example, the transition from pre-modern to Newtonian physics was a big leap in the level of systemicy (in the system of physical knowledge) and a giant reduction of complexity in the sense of information theory. Much fewer "bits" are required to describe Newton's world than that of Aristotle. The rise of systemicy — a rise in the degree of organization, one might say — in Western business enterprise over the past 150 years is not in itself a measure of rising complexity but rather a simplifying response to the increasing complexity of business environments. So while some authors think that modern technology is both becoming more complex and more systemic, this is an ambiguous and possibly misleading proposition.

The point of the conceptual scheme proposed is that it helps answer my first question: what is systemic about anything, including technology? By defining basic dimensions of systemicy, it provides a platform for developing measures of systemicy and thus for comparison between systems — from heaps of dust to human brains and telecommunication systems — in terms of degrees of systemicy.

Now, let me show how one could apply this concept of systemicy to technology and more particularly to the theory that modern technology is becoming increasingly systemic. The general suggestion is that technology is growing increasingly differentiated, inclusive, hierarchical and coherent. This, however, means very little until we specify what technology is. Bypassing any attempt to properly define this notoriously messy concept, I will simply state that, whatever "technology" may denote, it also connotes the domain of instrumentally operative principles of human activity. These operative principles appear in two basic forms: first, in an abstract form as a field of knowledge of instrumentally operative principles centered around methods, technical rules, and designs, and guided by norms of efficient operation. Secondly, operative principles appear in a socially embodied form, in the operative properties of instrumentally perceived material artifacts, in the routine operative skills of people, and in the organization of groups pursuing various operations collectively.

This notion of technology centers on operations. So if technology turns more systemic it means that technical operations become increasingly differentiated, inclusive, hierarchical and coherent. When applied to functionally specific operative units such as types of machinery or material systems of production and distribution, it is quite clear that many such units have grown more systemic in all four dimensions in the modern era. When applied to technology in general it is not so easy to say, not only because it is uncertain what one refers to but also because one could expect that changes in differentiation and inclusiveness are much easier to operationalize on the aggregate level than changes in hierarchy and coherence.

From Spencer's point of view, it may not even make much sense to apply systemicy to technology as a whole since his concept of "determination" referred to units of biological and cultural change, not to totalities. On the other hand, it does make sense to ask questions about technology viewed as a cultural sub-system, i.e. whether it grows more systemic, more interdependent, and more autodynamic.

Possibly one can pass by these difficulties by aggregating technology in terms of the environment for "typical" actions. In this sense, rising systemicy in technology implies that a typical modern act is conditioned by a technical environment displaying higher levels of operative differentiation, a greater volume of interactive technical elements, more hierarchical levels of control and a higher degree of dependence on the technical operations of other agents and systems. Thus, increasing technical systemicy in modern environments calls for higher degrees of specificity and fine tuning in typical individual action, revealing a higher degree of dependence on the restrictions and opportunities, as well as on cooperative behavior, of an increasingly inclusive and versatile technical environment.

Systemic Operations

Having outlined a conceptual platform for approaching systemicy I will now turn to my second question: what is at the root of systemicy in technology? What is it in technology that "interdepends" such that it acquires systemicy? Again the answer depends on what you mean by technology.

The roughly stated definition of technology above, centering on operations, provides a starting point. It suggests several directions in which to look for sources of systemicy, all of which focus on the interdependence of operations and operative characteristics.

One approach has been forwarded by the "bottleneck" theory and the notion of "technical imbalance" in production systems. The idea has been used by historians of technical and industrial change, e.g. by Lilley in his well-known *Men, Machines and History*.[4] More recently Rosenberg has contributed a theoretical elaboration of the approach.[5] To perform desired functions, the various operations and operants making up a technical system must fit internally. In individual machines this is a requirement that, at one end, conditions whether they will work at all, and at the other end, conditions their degree of efficiency. Fundations and gigs in machines must fit the load and speed of the transforming operations. Steering gadgets must fit the speed and latitude of cutting operations. All kinds of receivers must fit the operative properties of senders. Applied to systems of production, technical balance conditions efficiency and output possibilities. Operative capacities of spinning in textile production must fit capacities of weaving. Expanding the overall capacities of production systems demands countless innovative modifications of its components for it to stay internally fit. During expansion, perceived bottlenecks will direct innovative attention to inadequate operative components. Solutions to the problem of balancing operative properties against each other typically give rise to new imbalances elsewhere in the system, either in the form of further required adaptions or in the form of unexploited margins for improvements in operative efficiency or capacity. In this way any operative system, above a minimal level of systemicy, will acquire autodynamic properties in the sense that inducements in and directions for innovative technical change will be generated internally. Technical change breeds technical change. The desire to exploit functional gains from interoperative adaptions will typically serve to increase the level of systemicy in production systems, thus reinforcing autodynamics in them by increasing the volume of required and/or permitted internal adaptions.

A second approach has been forwarded by Hughes through his concept "reverse salient."[6] It is, as I see it, essentially an extension of the bottleneck theory since it includes more types of social structures into the notion of an operative system. Expansion in a production system not only meets with the primarily technical adaptive pressures highlighted by the bottleneck theory, but it also meets with many other types of adaptive requirements conditioning overall operative success — financial, organizational, political, and cultural. As a consequence, the analysis of change in production systems becomes more complicated as the horizon of operatively interactive elements widens. Another consequence is that the fairly strong element of autodynamics in technology, underlined in the "technical imbalance" approach, is played down, while the dependence of technical change on the wider cultural context is played "up".

A third approach to the problem of where to locate systemic operative interdependencies in technology would be to focus on their various social forms of appearance. As indicated above, operations appear both in a socially abstract form as a kind of knowledge — of methods, operators and designs — held by people, registered in documents, and organized in bodies of technical knowledge, research and education. On the other hand, operations are embodied in material artifacts, the skills of people, and in the rules and strategies of organizations pursuing the operations. This multiplicity of social forms represents in itself a kind of system with almost self-evident interdependencies and dynamic potential. Embodied technology is not only an application of abstract technology, but also formed by the interplay between different types of embodiment. The incessant traffic of operations between the operative skills of people and the analogous operative skills of machines and tools is probably the most important dimension of this. But regardless of how one stresses directions and priorities in the adaptive flows between different social forms of technology, it is generally credible that such interactions serve as a major source of systemic interdependence and dynamics in technology.

Now to a fourth line of thought. It rests on an attempt to abstract general kinds of operations not in terms of their social forms, but in terms of what they do. This provides another platform for conceptualizing systemicy in technology.

Though the number of functions or purposes of technology is limited only by human ingenuity and desire, the basic types of operation are few. These types of operations can be modelled in three dimensions. (1) Whatever result you intend to bring about with technology, and on whatever object (energy, material, information, and people) that technology operates, any operation either changes the states of these objects or keeps them from changing. From this perspective there are only two types of operations: changing and keeping. (2) Regardless of whether you try to change something or keep it from changing, your operation affects either the organization or the localization of the object in question: how it is organized and where it is in timespace. From this point of view, there are only two types of operations: organizing and localizing. (3) Every operation is either a first-order operation working directly on an object or it is of a higher-order operating, in the sense of controlling, another operation. Combining these options generates the following "three-dimensional" map:

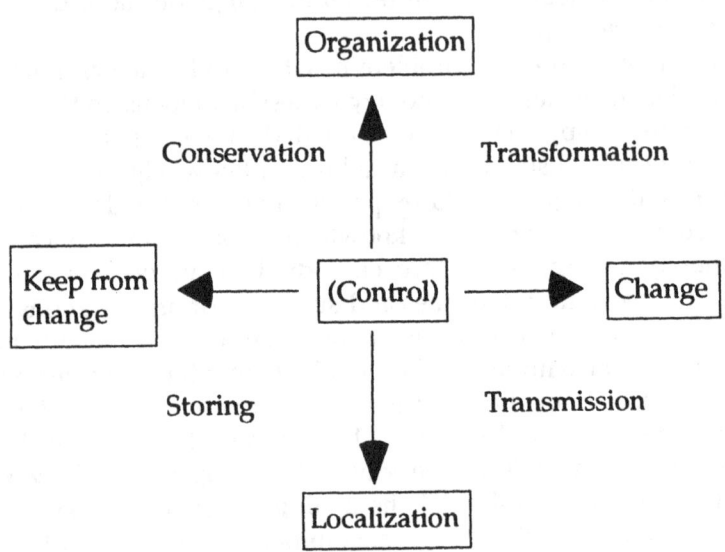

FIGURE 2 Basic types of system operations.

The general proposition is thus that any operation involves one or more of these basic types of operations. "Control" in the middle of the map indicates the third dimension, referring to operations that guide, restrict, activate, steer, and survey other operations and operators. Controlling is done by transforming, transmitting, storing and conserving something — as, for instance, a steering wheel essentially is an energy transmitter tied to a mechanism for transforming wheel angles. Control operations are thus only higher-level applications of the four fundamental operations defined by the map. (This explains why I have not bothered to depict all three dimensions graphically.)

Applying the categories of Figure 2 to technology can be done in several ways. One way is to characterize the overall operation of a technical object. So, a pair of scissors is mainly a transformator of materials, a refrigerator is mainly a conservator of decaying organic stuff, a book is mainly a transmittable store of information, and an automobile is mainly a transmitter of people and things. The overall operation typically depends on an ensemble of other operations sustaining it. To be a conservator of food, a refrigerator must also work as an energy transformator, an energy transmitter and a storeroom of cold air. An automobile is a configuration of hundreds of operators of all four kinds that contribute in various ways to the overall operation of moving things and people around.

Few, if any, useful technical things can be said to provide only one type of operation. One argument that "simple operators" in this sense do not exist even in principle is that every piece of material technology unintentionally stores, conserves and potentially transmits information on its own operative properties, and that every piece of material technology must, regardless of its operational characteristics, always have a substratum that stores and at least minimally conserves these operative properties. That is part and parcel of being materially embodied. Such necessary complexity is, however, rather trivial when discussing systemic technology, though, for example, the unintended self-informative properties of every piece of technology is certainly not trivial, when discussing the spread of technical knowledge.

Analyzing the interdependence between the four types of operations, including the levels of control to which they apply, allows for the identification of both micro- and macrolevel adaptive pressures of the "technical imbalance" type advocated by Rosenberg. If, following Hughes, one includes a wider spectrum of cultural components in the conception of an operative system, it may also be applied to the dynamics of "reverse salient." After all, it is partly because cultural facts control operations in the four areas of Figure 2 above that they become integral parts of the analysis of changes in technology. The addition of institutionally embodied levels of operative control is indeed a central feature of the emergence of large technical systems, signifying both their size and their systemicy. It is thus not a privilege for "technical facts" to have interdependent operative properties. Social institutions, like families, store people, conserve social relations, transform babies into citizens and transmit rules and resources between generations.

An analysis of operational dynamics in technology akin to this has already been suggested in the works of Rosenberg, Hughes and others. The tool suggested does not really differ from these authors' type of analysis, but it provides the analysis with an element of general conceptual rationalization. For example, any study of water-turbine electricity generation will reveal the powerful dynamics of a necessary fit between storing capacities of water, transmission capacities of pipes, transforming capacities of turbine generators, and conserving capacities of materials in dams and gigs. Likewise, any study of arms technology will reveal the extremely creative interaction between the operational features of attack and defense systems.

Rather than looking within the field of LTS studies for illustrations of how to apply the form of analysis suggested, I have chosen to illustrate it on a very grand scale: a speculative sketch of Western technological culture. The story starts with the revolutionary wave of storing innovation in the river delta cultures of Egypt and Mesopotamia about 5,000 years ago (water storing in dams, grain storing in stone houses, information storing in scriptures, value storing in metals, and power storing in cities and pyramids). This giant leap in storing capacity required and permitted adaptive changes in other types of operations. The most crucial was the amassing of local stores of wealth, which provided the base for a radical increase in trading (and war) opportunities. The epoch of leading storing technologies gave way to an Hellenistic-Roman-Arab epoch of trade (and robbery) when transport technologies came to the forefront as the prime movers of empires (ships, roads, canals, horse transportation, artillery, archery, aqueducts, and postal service). The rising volume of trade eventually generated pressures on local supplies and opportunities to increase local production, while at the same time rising transport capacities boosted risks and opportunities for warfare. Macrolevel innovative attention moved — we are now in the Middle Ages — to the problem of increasing transformation capacities in production and warfare. The modern technological era characterized by spectacular innovations in the transformation of materials and energies starts. These epochal changes in the focus of operations are also apparent in successive additions of functions in the master social technology of Western culture — money. Its first function was a store of value. Its second function was the transmitting of value as a medium of exchange. Its third added function was capital — a means of transforming values.

To make the move of Western history full circle through Figure 2 above, we can in our own times observe how the exceptional rise of cultural production and the destructive power of arms, following from the epoch of transformation, has created mounting problems for overall systems' stability. On the one hand, this makes conservation (of environments, information, energy, culture, and city life) come to the forefront of innovative imperatives. On the other hand, it imposes a general demand for control and increased operational hierarchy in order to permit giant volumes of production and to keep the Schumpeterian gale of creative destruction from shattering the social fabric entirely. The emergence of giant corporations and state apparatuses and, not least, the modern master technology of control — the computer — bear witness of this refocusing on operative imperatives. Consequently, we can also note the emergence of a fourth main function of money as an instrument of economic policy geared towards the conservation and control of national economic systems.[7]

This somewhat Hegelian example of the systems dynamics of interdependent basic operations on the level of world history should perhaps not be taken too seriously but I suggest that the principle involved should.

Systems Regulation and Its Alternatives

The basic business of the social sciences is the explanation of order and change in the social world. The history of social theory shows that this can be done in many ways and there is still little agreement on how it should be done properly. Considering systems regulation as one of several candidates for doing this job, or at least contributing to it, one must find a way to define systems regulation in the context of its principal alternatives. That is what I will try to do in this final section. With the help of a map of basic conceptions of social regulation, I will also provide an instrument for identifying major conflicting attitudes in the explanation of technical change.

Though there are indeed many social theories, those with a more general scope rest on a conception of the general nature of social regulation. When such conceptions are not stated explicitly, they are revealed by the patterns of explanation used. What is revealed in a particular theory is typically not one single principle of social regulation but several. Much of the basic uncertainty over the nature of social regulation seems to boil down to two overriding conceptual choices: agency vs. structure and program vs. process.[8]

Let us first look at agency vs. structure. Are order and change essentially products of human agency and thus the imprint of human intentionality, or is human action "structured" by super-individual social mechanisms such as traditions, institutions and markets? Is man the captain of history or rather the impotent crew of super-individual cultural vessels? Traditional social thought stressed agency, including the spectacular agency of gods, and it has only been in the last 200 years that the idea of social structure as a regulative principle has taken hold. Efforts to manage this antinomy, reeking with old perplexing issues of determinism and freedom, has occupied many master sociologists from Hegel on. Managing it is necessary not least because most social philosophers seem to think that the social world is regulated both by agency and by structure.

The second choice is program vs. process. In accounting for order and change in society should we be looking for some kind of program (intentions, rules, instincts, values, designs, norms, and ideologies) underlying and logically preceding them? Or is social order rather the generally unintended, unpreprogrammed outcome of a process in which a myriad of diverse human behaviors come to interact? Traditional social thought leaned heavily towards the program side, particularly in the form of explanations by "power." Much of the work of sociologists and anthropologists in the last two centuries has been devoted to identifying the regulative programs of social institutions. Functionalism in social thought has tried to spot hidden programs inherent in social structures. While the program approach has a very long history, it has only been since the end of the 18th century that a systematic grasp of various principles of process regulation in society — such as the market mechanism — has been taken.

Combining these two choices allows us to conceptualize four major types of regulative principles:

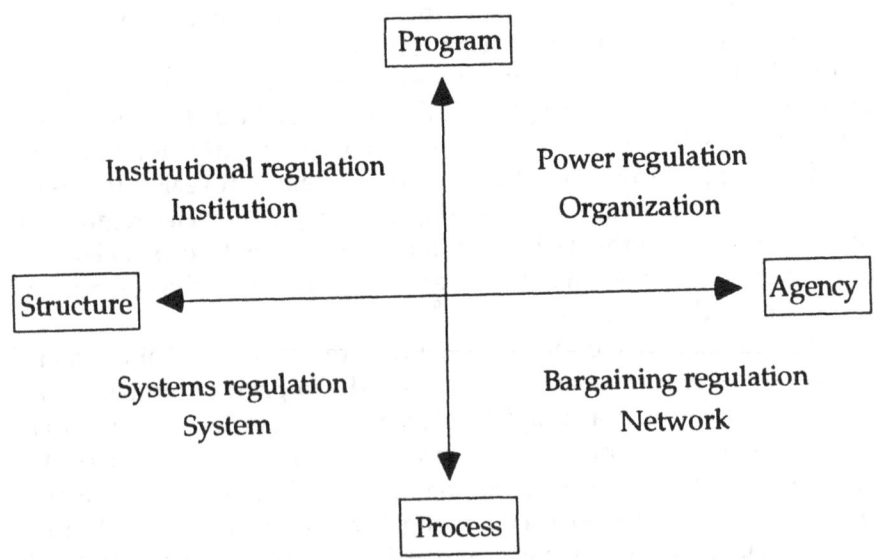

FIGURE 3 Basic forms of social regulation.

Combining "agency" with "program" gives the principle of power regulation: order and changes are accounted for by pointing to agents imposing their designs and intentions on the social world. In general organizational theory, this position explains the behavior of organizations as if they were rational agents themselves or fully controlled by human agents, typically contrasted with the conception of organization as a system — the diagonal alternative.[9] The power paradigm is the mainstream sociological attitude of people who are not social scientists, favored by journalists and business people, and heavily entailed in the very idea of politics as an activity of general social programing.

Combining "structure" with "program" gives the principle of institutional regulation. To account for social order and the patterns of human behavior, one should identify the institutionalized programs (of organizations, cultures, traditions, religions, ideologies, and other artifacts) that underlie it. By extending the notion of an institutional program, one can also include explanations of regulation in terms of functionally required prescriptions inherent in social structures and artifacts. According to Ralph Dahrendorf, this has been the mainstream of 20th century professional sociology ("norm-role sociology," "functionalism," and varieties of "structuralism").[10]

Combining "structure" with "process" gives the principle of systems regulation, of which the theoretically best articulated example is neo-classical, general equilibrium price theory (or general market theory). In such a perspective, the behavioral patterns of each individual is in principle accounted for through the relation to the behavior of all other agents. Nobody designs the resulting order and every rational agent is truly a prisoner of the system in the sense that he can, if fully informed, only choose that single action which happens to fit the conditions of interaction presented to him from the outside. The principal battle of modern social science has raged between the institutional paradigm of mainstream sociologists and the systems or market paradigm of mainstream economists. Even more alien to the systems paradigm of neo-classical economics is the power paradigm, reflected in the perennial tensions, moral and theoretical, between "politics" and "markets," as well as in the tendency of economists to view all forms of power regulation as something that obstructs the harmony of markets.[11] Apart from neo-classical economics, there are several other attempts to develop a general perspective on social change focusing on systems regulation. Some of these are found within the tradition of evolutionary theory.

Finally, combining "process" with "agency" gives the principle of bargaining regulation underlying approaches such as "conflict theory," "actor network theory," "bargaining theory," or "coalition theory." As can be expected from its position on the map, this perspective shares traits with both the power paradigm and the systems paradigm. "Bargaining" stresses that order is produced by free agency, while it underlines that such order does not result from particular designers but from the interaction of competing designers. The notion of a network stresses, on the one hand, the systemic interdependence of human action, while on the other, that there are no such universal frictionless arenas of interactions as in the neo-classical theory of markets. The process of interactive adaption is a fragmented business depending on the unstable structure of these networks. Most alien to the bargaining paradigm is the institutional paradigm. Consequently the recently renewed interest in bargaining regulation, evidenced in the social constructivist movement, runs parallel to an increasingly hostile attitude towards various forms of structuralism. Attempts to generalize and abstract principles of bargaining regulation are found within the general theory of games.

Each paradigm of regulation exposed in Figure 3 above can be associated with a general approach to the analysis of technological change. As in other fields of social theory, it is mostly a matter of which regulative principle to emphasize rather than an exclusive reliance on one type of regulation. The classical heroic inventor attitude — technology changes because great men conceive great things — is clearly a case of power regulation with an extreme emphasis on free agency and pre-programing. The equally heroic entrepreneur in Schumpeterian theories of innovation is another example. Many political strategies for technological innovation have a natural base in the power paradigm on the assumption that technical change can be successfully programmed by government agencies. When one accounts for technical change in terms of the strategies of domination pursued by giant corporations, one is also mainly working within this paradigm.

Theories of technical change emphasizing "technological paradigms," "trajectories," "guide-posts," and "regimes" have enjoyed popularity in recent years. Partly inspired by Kuhn's influential treatment of bodies of scientific theory as a kind of institution, the basic idea is to treat technology as a social institution which programs the behavior of technicians and the course of technical change for as long as they stand.[12] In accounting for the breakdown of such technical institutions, one has, like Kuhn, to change to another principle of regulation.

The systems paradigm of technological change stresses the dynamic adaption of routines, operational properties, and knowledge and skills in any part of a conceived technological system to changes in any other part of it. Characteristically it explains changes in technology through changes in technology. The theories of technical imbalance and bottlenecks, mentioned above, are developed mainly within this framework. The current rise of neo-evolutionary theory in innovation studies also belongs here.

Recent academic success of the bargaining paradigm in technology studies in the form of social constructivism represents a fourth master perspective on technical change.[13] It is apparently the relative novelty of a serious sociological interest in technology that accounts for the seemingly remarkable fact that all the master paradigms of regulation, as well as less easily discernible mixes of them, appear in technology studies at the same time. On the assumption that all four principles represent real regulative mechanisms in parallel, interacting work in most forms of social change, it is not so remarkable that you find them also in technology studies.

On that assumption, the problem is not one of choosing which of these regulative principles to trust exclusively. It is rather the problem of showing how they interact in a manner that does not succumb to threadbare eclecticism. This problem is typically tackled by combining regulative principles in a particular way. Here we have theories such as Weber's idea about an oscillation in history between longer periods of institutional regulation, interfoliated by shorter creative periods of power regulation. It is overshadowed in the modern world by an increasingly important principle of systems regulation, i.e. his theory of increasing cultural rationalization. We have theories such as Schumpeter's idea about an oscillation between innovative, entrepreneurial, power regulation and reproductive systems regulation by the market. Popper's influential theory of scientific change is essentially of this kind too, featuring a dynamic interplay between bold conjectures and the systemic adaptions within the stock of knowledge. There is also the "social closure theory" of some social constructivists, inspired by Weber and Kuhn, about an oscillation between the "open" bargaining paradigm valid for early phases in the emergence of a technology and the "closed" institutional paradigm valid for the interpretation of established technologies. Interaction over the bargaining-institution diagonal in Figure 3 above is what we find in several classics, such as Hegel and Marx. We have, not least, the recent heroic attempt of Giddens' to combine all four regulative principles in his "structuration theory."[14]

The general problem in all such efforts to combine several regulative principles is to show how they interact to produce social order and change. The specific problem is to explain changes in the composition of regulative principles and to account for variations and trends in the pattern of regulation.

That these principles generally apply to current theories of technical change has already been suggested. But how do they relate to the notion of system and to the theory that technology is growing increasingly systemic?

One way to approach this question is to ask whether calling something a "large technical system" implies that it is mainly systems regulated in the sense of Figure 3. As it seems, this is not generally the case. In fact it may be disputed whether large technical systems should be considered as systems regulated by the systems paradigm, as organizations regulated by the power paradigm, as networks regulated by the bargaining paradigm, or as institutions regulated by the institutional paradigm. Descriptions of various large technical systems may obviously differ in this respect, depending not only on theoretical preferences but also on historically specific conditions of the properties and contexts of such systems. Most conspicuously, these units are regulated differently at different stages of their development. Of particular interest are thus descriptions suggesting basic changes in the mode of regulation of these systems. Some of the radical reconfigurations highlighted in this volume may illustrate this. Several LTS in electricity, telecommunications, and transportation have recently been affected by a wave of deregulation. This implies a general drift to the bottom half of figure 3. Program regulation gives way to process regulation. Partly, this change is in itself program regulated in the sense that it reflects a widespread renaissance for political trust in the venerable "invisible hand," i.e. market regulation. Partly, it reflects the consequence of a technologically determined destruction of barriers to interaction and entry in regional systems. One may also assume that world-wide computerization represents such dramatic changes in the conditions of control and coordination as well as a general decrease of stability in technical environments that established forms of program regulation are weakened.

Finally, the question is whether theories of increasing systemicy in technology entail that systems regulation is growing relatively more important in technical change. At first glance, this is a credible proposition, particularly if growing systemicy in technology is coupled to the idea of an increasing autonomy of technical change. The systems regulation paradigm is the natural stronghold for all kinds of "internalism" in social change theory. Though I believe that the most reasonable way to argue in favor of the autonomous technology thesis would be to demonstrate increasing systemicy in technology in a sense entailing an increasing share for systems regulation, it is not the only way. An alternative would be the demonstration of a general trend towards process regulation in technical change, regardless if this goes towards increasing systems regulation or increasing bargaining regulation. Both imply that the margin for programing technical change is shrinking. If autonomy in technology primarily means that technical change cannot be effectively programmed "by us," perhaps in the sense of democratic governments, then a preponderance of process regulation in technical change entails its autonomy.

Another alternative is suggested in the thoughts of Ellul. One of his main arguments concerning autonomous technology is framed by his concept "la technique." This is an institutional concept connoting a cultural frame of mind, a general program for social action focusing on instrumental efficiency and guided by a vision of "the one best way." Technology becomes autonomous in the sense that this institutional program dominates over other institutionalized forms of social reason.

There is, finally, the possibility to interpret the notion of systemicy in technology in a way that is not specifically tied to systems regulations but to structural regulation in general, i.e. going to the left in figure 3. Being systemic may then connote a tightly knit institutional web, keeping agents in something like Weber's famous "iron cage." Autonomy in technical change can thus be argued for in terms of any kind of move from "agency" to "structure" in the mode of regulation.

Conclusion

So, to conclude, there is neither a necessary connection between being called a technical system and being systems regulated. Nor is there a necessary connection between systems regulation and ideas of autonomy in technological change. There is not even a necessary connection between being systemic and being systems regulated. This certainly calls for some caution in the suggestive use of "system" in technology studies.

In this essay I have raised three general questions about systemic technology. In order to answer them I have devised three conceptual tools and suggested how they may be useful in developing proper answers. If they actually are useful remains for the reader to judge.

Acknowledgements

I would like to thank Jenny Beckman, Bernwald Joerges and Jane Summerton for helpful comments on earlier versions of this paper.

Notes

1. Evidence of this is the absence of an established vocabulary. Terms like systemic, systemicy (or systemicity) and systemization (for the process of growing increasingly systemic) have only recently entered scientific jargon. They are not as yet listed in major dictionaries.
2. Hannay and McGinn, 1980; Winner, 1987.
3. Spencer, 1970.
4. Lilley, 1965.
5. Rosenberg, 1976, pp. 108-125.
6. Hughes, 1983, p. 46. In favor of "reverse salient", Hughes argues that the bottleneck approach applies only to mechanical technical systems. I am not convinced that this is necessarily so.
7. For an earlier version of this speculation see Beckman, 1987.
8. The argument is exposed in more detail in Beckman, 1990, chap. 7.
9. For an overview of this controversy see Abrahamsson, 1993.
10. Dahrendorff, 1973.
11. Lindblom, 1977.
12. Kuhn, 1963, Dosi, 1984; Sahal, 1981.
13. Bijker, Hughes and Pinch, eds., 1989, MacKenzie and Vajcman, 1985.
14. Scranton, 1993.

References

Abrahamsson, Bengt. 1993. *The Logic of Organizations*. London: Sage Publication, Inc.

Beckman, Svante et. al. 1987. *Rabalder i människans provins: Fem forskare om datasamhället*. Linköping University.

Beckman, Svante. 1987. "Datorer i världshistorisk belysning." In *Rabalder i människans provis: Fem forskare om datasamhället*, ed. Beckman, Svante, et.al. Linköping University.

_____. 1990. *Utvecklingens hjältar: Den innovativa individen i samhällstänkandet*. Stockholm: Carlsson förlag.

Bijker, Wiebe, Thomas P. Hughes and Trevor Pinch, eds. 1987. *The Social Construction of Technology: New Directions in the Sociology and History of Technology*. Cambridge, Mass: The MIT Press.

Dahrendorff, Ralph. 1973. *Homo Sociologicus*. London: Routhledge & Kegan Paul.

Dosi, Giovanni. 1984. *Technical Change and Industrial Transformation*. New York: Macmillan.

Hannay, N. Bruce and Robert E. McGinn. 1980. "The Anatomy of Modern Technology." *Daedalus* 109, no. 1: 25-53.

Hughes, Thomas P. 1983. *Networks of Power: Electrification in Western Society 1880 — 1930*. Baltimore: Johns Hopkins University Press.

Kuhn, Thomas. 1963. *The Structure of Scientific Revolutions.* Chicago: University of Chicago Press.

Lilley, Samuel. 1965. *Men, Machines and History.* New York: International Publishers.

Lindblom, Charles E. 1977. *Politics and Markets: The World Political-Economic System.* New York: Basic Books.

MacKenzie, Donald and Judy Vajcman. 1985. *The Social Shaping of Technology: How the Refrigerator Got Its Hum.* Philadelphia: Open University Press.

Rosenberg, Nathan. 1976. *Perspectives on Technology.* London: Cambridge University Press.

Sahal, Dieter. 1981. "Alternative Conceptions of Technology." *Research Policy* 10: 2 — 24.

Scranton, Philip. 1993. "Giddens' Structuration Theory and Research on Technical Change". Workshop paper given at the Conference on Technical Change at Oxford University, 8-10 September 1993.

Spencer, Herbert. 1970. *Social Statics.* New York: Robert Schalkenbach Foundation.

Winner, Langdon. 1977. *Autonomous Technology: Technics-Out-Of-Control as a Theme in Political Thought.* Cambridge, Mass: The MIT Press.

About the Contributors

Janet Abbate is a historian and doctoral student in the Department of American Civilization at the University of Pennslyvania. She is currently finishing her dissertation on the history of the ARPANET and Internet computer networks as well as doing research on policy issues in information infrastructure.

Svante Beckman is an economic historian and associate professor at the Department of Technology and Social Change, Linköping University (Sweden). His publications include works on technology and power, professionalization and theories of social change. He is currently writing a book on the concept of technical progress.

Ingo Braun is currently a fellow of the Rathenau Foundation for the History of Science. He has worked in the "large technical systems" group at Wissenschaftszentrum Berlin für Sozialforschung. He has written *Stoff-wechsel-Technik. Zur Soziologie und Ökologie der Waschmaschinen,* (1988);*Technik-Spiralen. Vergleichende Studien zur Technik im Alltag* (1993) and co-edited *Technik ohne Grenzen* (with Bernward Joerges, forthcoming 1994). His current research has two foci: the worldwide development of Internet and the scientific career of garbage.

Arden Bucholz is professor of history and director of the graduate program in history at State University of New York, College at Brockport. His published works include *Hans Delbruck and the German Military Establishment* (1985) and *Moltke Schlieffen and Prussian War Planning* (1991). He is currently working on a book on European war planning 1871-1914 for which the essay in this volume is an outline.

Olivier Coutard is an economist and researcher at the Centre National de la Recherche Scientifique in Noisy-le-Grand, France. He has co-authored "Information et gestion dynamique, ou quand les réseaux deviennent intelligents" (1990, "Information and dynamic management, or when the networks become intelligent.")

Reiner Grundmann holds a Ph.D. in social and political sciences. He has been a fellow at the Rathenau Foundation for the History of Science as well as Wissenschaftszentrum Berlin für Sozialforschung. His publications include *Marxism and Ecology* (1991). He is presently working on risk regulation, specifically the history of the ozone hole controversy and its political regulation, at the University of Bremen.

Thomas P. Hughes is Mellon professor of history and sociology of science at the University of Pennsylvania. His many published works include *Elmer Sperry: Inventor and Engineer* (hardcover 1971, softcover 1993), *Networks of Power: Electrification of Western Society 1880-1930* (hardcover 1983, softcover 1993), and *American Genesis: A Century of Invention and Technological Enthusiasm 1870 - 1970* (1990). He has also co-edited (with Agatha Hughes) *Lewis Mumford: Public Intellectual* (1990).

Bernward Joerges is a Senior Research Fellow at the Wissenschaftszentrum Berlin für Sozialforschung, where he heads the Metropolitan Research Group, and professor of sociology at the Technical University Berlin. He has published widely on technology in everyday life (for example *Technik im Alltag*, 1988) as well as large technical systems (*Technik ohne Grenzen*, co-edited with Ingo Braun, forthcoming 1994). He is currently working on big city infrastructure and management.

Oskar Juhlin is an engineer and doctoral student at the Department of Technology and Social Change, Linköping University. He has published articles on engineering practices from a constructivist perspective and is currently working on a micro-sociological study of engineering practices related to the development of so-called Road Transport Informatics.

Tobias Robischon is a political scientist and scholar at the Graduiertenkolleg für Sozialwissenschaften, Cologne University, and is also affiliated with the Max-Planck-Institut für Gesellschaftsforschung. He has previously written on how to shape computer networks to fit the needs of new social movements. He is currently working on his dissertation on the transformation of telecommunications systems in the former German Democratic Republic.

Gene I. Rochlin is professor in the Energy and Resources Group and research policy analyst at the Institute of Governmental Studies, University of California at Berkeley. His published works include *Science, Technology, and Social Change* (1974) and *Plutonium, Power and Politics: International Arrangements for the Disposition of Spent Nuclear Fuel* (1979). He is currently completing a book on the social and organizational consequences of the functional integration of computers into a wide range of human activities.

Stephen M. Salsbury is professor of economic history and dean of the faculty of economics at the University of Sydney. He has written *The State, the Investor, and the Railroad: The Boston & Albany 1825-1867* (1967) and *No Way to Run a Railroad: The Untold Story of the Penn Central Crisis* (1982). He has also co-authored *Pierre S. Du Pont and* *the Making of the Modern Corporation* (1971); *The Bull, the Bear and the Kangaroos: A History of the Sydney Stock Exchange* (1988); and *Sydney Stockbrokers: Biographies of Members of the Sydney Stock Exchange 1871 - 1987* (1992). He is currently writing a history of one of Australia's leading commercial banks (Westpac).

Volker Schneider is a political scientist and research fellow at the Max-Planck-Institut für Gesellschaftsforschung in Cologne. His published works include *Politiknetzwerke der Chemikalienkontrolle* (1988, on political networks for chemical control) and *Technikentwicklung zwischen Politik und Markt: Der Fall Bildschirmtext* (1989, on politics and market in technology development as seen in computer screen texts). He is currently working on technology studies and policy networks.

Jane Summerton holds a Ph.D. in technology and social change. She has been a Visiting Scholar at the University of California at Berkeley and is currently a research associate in the Department of Technology and Social Change, Linköping University. Her published works include *District Heating Comes to Town: the Social Shaping of an Energy System* (1992). She is currently studying institutional change and actors' strategies in the reshaping of electricity systems in Sweden, Norway and parts of the United States.

Steven W. Usselman is associate professor of history at the University of North Carolina at Charlotte. He has published numerous articles on innovation in the American railroad industry in journals such as *Business History Review* and *Technology and Culture*. He is currently completing a book entitled *Regulating Innovation: The Management, Economics, and Politics of Technical Change in American Railroads, 1860-1914*. His other research interests include the international computer industry.

Alexandra von Meier is a doctoral student in Energy and Resources at the University of California, Berkeley. She has written on photovoltaics and their economic evaluation, hydrogen energy systems, and operational culture at nuclear power plants. Her current work is on cultural factors in technology decisions, using a case study of distribution automation in electric utilities.

About the Book

Because of their important roles in contemporary society, large technical systems—such as railways, airlines, road systems, telecommunications, and electric power networks—are currently attracting considerable academic and political interest. In this international anthology on processes of change in large technical systems, the contributing authors present historical and current case studies of transformation within these systems.

Working at the forefront of historical and social science research on the dynamics of large technical systems, the authors specifically analyze how and why such systems undergo change. In some cases, new technologies are solving old problems and presenting opportunities for system growth. In other areas, new regulatory approaches have brought competition and deregulation, often posing complex challenges to system builders. The authors also show how the breakup of national boundaries and new corporate strategies for global management of technology are transforming systems in ways that will have significant impacts on all consumers.

Index

"This is an incredibly timely and useful resource, one that will be helpful for the novice and seasoned practitioner and supervisor alike. Baker and Cross address assessment, treatment, and prevention to deal with the obvious and hidden signs of suicidal thinking, preoccupation, and planning. Many of the chapters are written by Baker and Cross from a prescriptive play therapy lens. The other chapters feature renowned authors and cover racial and cultural differences and inclusion, age and gender, ethics, neurodivergence, psychiatry, countertransference, school settings and technology, and special issues utilizing expressive arts and play therapy approaches. This comprehensive book will not be a one-time read but rather a resource to return to again and again. For those of us on the front lines of treatment, as well as play therapy leaders teaching and mentoring the rising generations of play therapists, this is absolutely required reading."

—**Athena A. Drewes, Psy.D., MA, MS, Ed, RPT-S**, *Founder and President Emeritus of the New York Association for Play Therapy and former Director of the Association for Play Therapy*

"This much-needed text addresses one of the most challenging and detrimental issues faced by clinicians working with children and adolescents. It is theoretically sound, comprehensive, and practical. Breaking new ground in the field, this publication will serve the needs of both novice and seasoned clinicians as they work with one of our most vulnerable populations."

—**Sueann Kenney-Noziska, MSW, LCSW, RPT-S**, *Play Therapy Corner, Las Cruces, New Mexico*

"The searing pain of youth longing for connection in a world of pervasive disconnection fed by social media makes this book on the assessment, prevention, and treatment of suicidal behavior in youth essential. This book is comprehensive, pays close attention to cultural issues, and reflects the wisdom of both the editors and its distinguished contributors. Highly recommended!"

—**David A. Crenshaw, Ph.D., ABPP**, *author and Chief of Clinical Services at the Children's Home of Poughkeepsie*

"Cross and Baker offer mental health providers a map for approaching suicidal ideation in children and adolescents while demonstrating the therapeutic power of play as an effective treatment. I hope all mental health providers read this book to help destigmatize the topic and help ease the anxiety when working with this population. As Cross and Baker note in the book: 'Because you (the child) are worth caring about.' "

—**Liliana Baylon, LMFT-S, RPT-S**, *bilingual and bicultural therapist*

"This book fills an important gap in the mental health field, providing play therapists with a culturally inclusive look at how we can effectively address suicide in children from all backgrounds. The authors highlight the effects of the COVID-19 pandemic on all young people, with a special focus on marginalized children. In applying the therapeutic powers of play through expressive therapies, the authors are providing play therapists with the skills and competencies they need to engage children, caregivers, and school-based professionals in combatting the epidemic of suicide in young people."

—**April Duncan, DSW, LCSW, RPT-S™**, *Founder and CEO of BMH Connect*